Numerical Methods for the Solution of Ill-Posed Problems

Mathematics and Its Applications

Mathematics and Its Applications

Managing Editor:

M. HAZEWINKEL

Centre for Mathematics and Computer Science, Amsterdam, The Netherlands

Volume 328

Numerical Methods for the Solution of Ill-Posed Problems

by

A. N. Tikhonov[†]
A. V. Goncharsky
V. V. Stepanov
A. G. Yagola

Moscow State University,
Moscow, Russia

KLUWER ACADEMIC PUBLISHERS
DORDRECHT / BOSTON / LONDON

A C.I.P. Catalogue record for this book is available from the Library of Congress

ISBN 978-90-481-4583-6

Published by Kluwer Academic Publishers,
P.O. Box 17, 3300 AA Dordrecht, The Netherlands.

Kluwer Academic Publishers incorporates
the publishing programmes of
D. Reidel, Martinus Nijhoff, Dr W. Junk and MTP Press.

Sold and distributed in the U.S.A. and Canada
by Kluwer Academic Publishers,
101 Philip Drive, Norwell, MA 02061, U.S.A.

In all other countries, sold and distributed
by Kluwer Academic Publishers Group,
P.O. Box 322, 3300 AH Dordrecht, The Netherlands.

Printed on acid-free paper

This is a completely revised and updated translation of the Russian
original work *Numerical Methods for Solving Ill-Posed Problems*,
Nauka, Moscow © 1990
Translation by R.A.M. Hoksbergen

Contents

Preface to the English edition

The Russian original of the present book was published in Russia in 1990 and can nowadays be considered as a classical monograph on methods for solving ill-posed problems. Next to theoretical material, the book contains a FORTRAN program library which enables readers interested in practical applications to immediately turn to the processing of experimental data, without the need to do programming work themselves. In the book we consider linear ill-posed problems with or without a priori constraints. We have chosen Tikhonov's variational approach with choice of regularization parameter and the generalized discrepancy principle as the basic regularization methods. We have only fragmentarily considered generalizations to the nonlinear case, while we have not paid sufficient attention to the nowadays popular iterative regularization algorithms. A reader interested in these aspects we recommend the monograph: 'Nonlinear Ill-posed Problems' by A.N. Tikhonov, A.S. Leonov, A.G. Yagola (whose English translation will be published by Chapman & Hall) and 'Iterative methods for solving ill-posed problems' by A.B. Bakushinsky and A.V. Goncharsky (whose English translation has been published by Kluwer Acad. Publ. in 1994 as 'Ill-posed problems: Theory and applications'). To guide the readers to new publications concerning ill-posed problems, for this edition of our book we have prepared a Postscript, in which we have tried to list the most important monographs which, for obvious reasons, have not been included as references in the Russian edition. We have not striven for completeness in this list.

In October 1993 our teacher, and one of the greatest mathematicians of the XX century, Andreĭ Nikolaevich Tikhonov died. So, this publication is an expression of the deepest respect to the memory of the groundlayer of the theory of ill-posed problems. We thank Kluwer Academic Publishers for realizing this publication.

A.V. Goncharsky, V.V. Stepanov, A.G. Yagola

Introduction

From the point of view of modern mathematics, all problems can be classified as being either correctly posed or incorrectly posed.

Consider the operator equation

$$Az = u, \qquad z \in Z, \quad u \in U, \tag{0.1}$$

where Z and U are metric spaces. According to Hadamard [213], the problem (0.1) is said to be *correct*, *correctly posed* or (Hadamard) *well posed* if the following two conditions hold:

    a) for each $u \in U$ the equation (0.1) has a unique solution;

    b) the solution of (0.1) is stable under perturbation of the righthand side of this equation, i.e. the operator A^{-1} is defined on all of U and is continuous.

Otherwise the problem (0.1) is said to be *incorrectly posed* or *ill posed*.

A typical example of an ill-posed problem is that of a linear operator equation (0.1) with A a compact operator. As is well known, in this case both conditions for being Hadamard well posed can be violated. If Z is an infinite-dimensional space, then, first, A^{-1} need not be defined on all of U ($AZ \neq U$) and, secondly, A^{-1} (defined on $AZ \subset U$) need not be continuous.

Many problems from optimal control theory and linear algebra, the problem of summing Fourier series with imprecisely given coefficients, the problem of minimizing functionals, and many others can be regarded as ill-posed problems.

Following the publication of the ground-laying papers [164]–[166], the theory and methods for solving ill-posed problems underwent extensive development. The most important discovery was the introduction, in [166], of the notion of approximate solution of an ill-posed problem. The notion of *regularizing algorithm* (RA) as a means for approximately solving an ill-posed problem lies at the basis of a new treatment.

Consider the following abstract problem. We are given metric spaces X and Y and a map $G \colon X \to Y$ defined on a subset $D_G \subset X$. For an element $x \in D_G$ we have to find its image $G(x) \in Y$. Returning to the problem (0.1), in this new terminology we have $G = A^{-1}$, $X = U$, $Y = Z$, and the problem consists of computing A^{-1}. In this setting, $D_G = AZ \subset U$.

The map G may play, e.g., the role of operator mapping a set of data for some

1

extremal element of the problem into an element on which the extremum is attained, etc.

If G is defined on all of X and is continuous, the problem under consideration is Hadamard well posed. In this case, if instead of the element $\overline{x} \in D_G$ we know an approximate value of it, i.e. an element $x_\delta \in X$ such that $\rho(x_\delta, \overline{x}) \leq \delta$, then we can take $G(x_\delta) \in Y$ as approximate value of $\overline{y} = G(\overline{x})$; moreover, $\rho_Y(G(x_\delta), \overline{y}) \to 0$ as $\delta \to 0$.

If the problem is ill posed, then $G(x_\delta)$ need not exist at all, since x_δ need not belong to D_G, while if it does belong to D_G, then in general $\rho_Y(G(x_\delta), \overline{y})$ need not tend to zero as $\delta \to 0$.

Thus, the problem under consideration may be treated as the problem of approximately computing the value of an abstract function $G(x)$ for an imprecisely given argument x. The notion of *imprecisely given argument* needs a definition. An *approximate data* for $\overline{x}$ is understood to mean a pair (x_δ, δ) such that $\rho_X(x_\delta, \overline{x}) \leq \delta$, where x_δ is not required to belong to D_G.

Fundamental in the theory of solving ill-posed problems is the notion of regularizing algorithm as a means for approximately solving an ill-posed problem. Consider an operator R defined on pairs (x_δ, δ), $x_\delta \in X$, $0 < \delta \leq \delta_0$, with range in Y. We can use another notation for it, to wit: $R(x_\delta, \delta) = R_\delta(x_\delta)$, i.e. we will talk of a parametric family of operators $R_\delta(x)$, defined on all of X and with range in Y. Consider the discrepancy

$$\Delta(R_\delta, \delta, \overline{x}) = \sup_{\substack{x_\delta \in X \\ \rho_X(x_\delta, \overline{x}) \leq \delta}} \rho_Y\left(R_\delta(x_\delta), G(\overline{x})\right).$$

DEFINITION. The function G is called *regularizable on D_G* if there is a map $R(x, \delta) = R_\delta(x)$, acting on the direct product of the spaces X and $\{\delta\}$, such that

$$\lim_{\delta \to 0} \Delta(R_\delta, \delta, x) = 0$$

for all $x \in D_G$. The operator $R(x, \delta)$ $(R_\delta(x))$ itself is called a *regularizing operator* (*regularizing family of operators*).

The abstract setting of the problem, presented above, comprises various ill-posed problems (solution of operator equations, problem of minimizing functionals, etc.). The notion of regularizability can be extended to all such problems. For example, the problem (0.1) is regularizable if A^{-1} is regularizable on its domain of definition $AZ \subset U$. In this case there is an operator R mapping a pair $u_\delta \in U$ and $\delta > 0$ to an element $z_\delta = R(u_\delta, \delta)$ such that $z_\delta \xrightarrow{Z} \overline{z}$ as $\delta \to 0$.

In regularization theory it is essential that the approximation to $G(\overline{x})$ is constructed using the pair (x_δ, δ).

Clearly, when constructing an approximate solution we cannot use the triplet $(x_\delta, \delta, \overline{x})$, since the exact value of $\overline{x}$ is unknown. The problem arises whether it is possible to construct an approximate solution using x_δ only (the error δ being unknown, while it is known that $\rho_X(x_\delta, \overline{x}) \to 0$ as $\delta \to 0$). The following assertion shows that, in essence, this is possible only for problems that are well posed.

The map $G(x)$ is regularizable on D_G by the family $R_\delta = R(\cdot, \delta) = R(\cdot)$
if and only if $G(x)$ can be extended to all of X and this extension is
continuous from D_G to X (see [16]).

The latter means that in the ill-posed problem (0.1) (e.g., suppose that the operator A^{-1} in (0.1) is bijective from Z onto U and compact) the map $G = A^{-1}$ cannot be extended onto the set U with $D_G = AZ \subset U$ such that it would be continuous on D_G, since the operator A^{-1} is not continuous. This means that in the above mentioned regularization problem we cannot use an operator $R(\cdot)$ that is independent of δ.

Thus, the pair (x_δ, δ) is, in general, the minimal information necessary to construct an approximate solution for ill-posed problems. Correspondingly, the pair (x_δ, δ) represents minimal information in (0.1).

We pose the following question: how wide is the circle of problems that allow the construction of a regularizing family of maps, i.e. try to describe the circle of problems that are Tikhonov regularizable. It is clear that this set of problems is not empty, since for any correct problem we can take $R_\delta = G$ as regularizing family. In essence, all of classical numerical mathematics is based on this fact. Not only the well-posed problems, but also a significant number of ill-posed problems are regularizable. For example, if the operator A in (0.1) is linear, continuous and injective, and if Z and U are Hilbert spaces, the resulting problem is regularizable.

This result is the main result in Chapter 1. Moreover, in this first Chapter we propose constructive methods for obtaining regularizing algorithms for the problem (0.1) in case not only the righthand side but also the operator involves an error. Suppose we are given an element u_δ and a linear operator A_h such that $\|u_\delta - \overline{u}\|_U \leq \delta$, $\|A_h - A\| \leq h$. In other words, the initial information consists of $\{u_\delta, A_h, \delta, h\}$. We are required to construct from this data an element $z_\eta \in Z$, $\eta = \{\delta, h\}$ such that $z_\eta \xrightarrow{Z} \overline{z}$ as $\eta \to 0$. The following construction for solving this problem is widely used. Consider the functional

$$M^\alpha[z] = \|A_h z - u_\delta\|_U^2 + \alpha\|z\|_Z^2. \tag{0.2}$$

Let z_η^α be an extremal of the functional $M^\alpha[z]$, i.e. an element minimizing $M^\alpha[z]$ on Z. If the *regularization parameter* $\alpha = \alpha(\eta)$ matches in a certain sense the set $\eta = \{\delta, h\}$, then in a certain sense $z_\eta^{\alpha(\eta)}$ will be a solution of (0.1).

In Chapter 1 we give a detailed discussion of the means for matching the regularization parameter α with the specification error η of the initial information. We immediately note that trying to construct an approximate solution such that α is not a function of η is equivalent to trying to construct a regularizer $R(\cdot, \eta) = R(\cdot)$. As we have seen, this is possible only for well-posed problems.

In Chapter 1 we also give a detailed discussion of the a priori schemes for choosing the regularization parameter which were first introduced in [165].

Of special interest in practice are schemes for choosing the regularization parameter using generalized discrepancies [58], [185]. We also give a detailed account of methods for solving incompatible equations.

In Chapter 1, considerable space is given to problems of finite-difference approximation and to numerical methods for solving the system of linear equations obtained after approximation. Special attention is given to the modern methods for solving integral equations of convolution type. Regularizing algorithms for solving equations involving differences of arguments have acquired broad applications in problems of information processing, computer tomography, etc. [183], [184].

The construction of regularizing algorithms is based on the use of additional information regarding the solution looked for. This problem can be solved in a very simple manner if information is available regarding the membership of the required solution to a compact class [185]. As will be shown in Chapter 2, such information is fully sufficient in order to construct a regularizing algorithm.

Moreover, in this case there is the possibility of constructing not only an approximate solution z_δ of (0.1) such that $z_\delta \longrightarrow^Z \overline{z}$ as $\delta \to 0$, but also of obtaining an estimate of the accuracy of the approximation, i.e. to exhibit an $\epsilon(\delta)$ for which $\|z_\delta - \overline{z}\|_Z \leq \epsilon(\delta)$, where, moreover, $\epsilon(\delta) \to 0$ as $\delta \to 0$.

The problem of estimating the error of a solution of (0.1) is not simple.

Let

$$\Delta(R_\delta, \delta, \overline{z}) = \sup_{u_\delta:\, \|u_\delta - \overline{u}\| \leq \delta} \rho_Z\left(R_\delta(u_\delta), \overline{z}\right)$$

be the error of a solution of the ill-posed problem (0.1) at the point $\overline{z}$ using the algorithm R_δ. It turns out that if the problem (0.1) is regularizable by a continuous map R_δ and if there is an error estimate which is uniform on D,

$$\sup_{\overline{z} \in D} \Delta(R_\delta, \delta, \overline{z}) \leq \epsilon(\delta) \to 0,$$

then the restriction of A^{-1} to $AD \subset U$ is continuous on $AD \subset U$ [16]. This assertion does not make it possible to find the error of the solution of an ill-posed problem on all of Z.

However, if D is compact, then the inverse operator A^{-1}, defined on $AD \subset U$, is continuous. This can be used to find, together with an approximate solution, also its error.

The following important question with which Chapter 2 is concerned is as follows. For a concrete problem, indicate a compact set M of well posedness, given a priori information regarding the required solution.

In a large number of inverse problems of mathematical physics there is qualitative information regarding the required solution of the problem, such as monotonicity of the functions looked for, their convexity, etc.[71]. As will be shown in Chapter 2, such information suffices to construct a RA for solving the ill-posed problem (1.1) [74].

The next problem solved in Chapter 2 concerns the construction of efficient numerical algorithms for solving ill-posed problems on a given set M of well-posedness. In the cases mentioned above, the problem of constructing an approximate solution reduces to the solution of a convex programming problem. Using specific features of the constraints we can construct efficient algorithms for solving ill-posed problems on compact sets.

We have to note that although in this book we extensively use iteration methods to construct regularizing algorithms, we will not look at the problem of iterative regularization. This is an independent, very large field of regularization theory in itself. Several monographs have been devoted to this direction, including [16], [31].

In Chapter 3 we study in detail the numerical aspects of constructing efficient regularizing algorithms on special sets. Algorithms for solving ill-posed problems on sets of special kinds (using information regarding monotonicity of the required solution, its convexity, existence of a finite number of inflection points, etc.) have become widespread in the solution of the ill-posed problems of diagnostics and projection [51], [52], [183].

In Chapter 4 we give the description of a program library for solving ill-posed problems. This library includes:

a) various versions for solving linear integral equations of the first kind (0.1), based on the original Tikhonov scheme;

b) special programs for solving convolution-type one- and two-dimensional integral equations of the first kind, using the fast Fourier transform;

c) various versions for solving one-dimensional Fredholm integral equations of the first kind on the set of monotone, convex functions, on the sets of functions having a given number of extrema, inflection points, etc.

Each program is accompanied by test examples. The Appendix contains the program listings.

This book is intended for students with a physics-mathematics specialisation, as well as for engineers and researchers interested in problems of processing and interpreting experimental data.

CHAPTER 1

Regularization methods

In this Chapter we consider methods for solving ill-posed problems under the condition that the a priori information is, in general, insufficient in order to single out a compact set of well-posedness. The main ideas in this Chapter have been expressed in [165], [166]. We will consider the case when the operator is also given approximately, while the set of constraints of the problem is a closed convex set in a Hilbert space. The case when the operator is specified exactly and the case when constraints are absent (i.e. the set of constraints coincides with the whole space) are instances of the problem statement considered here.

1. Statement of the problem. The smoothing functional

Let Z, U be Hilbert spaces, D a closed convex set of a priori constraints of the problem $(D \subseteq Z)$ such that $0 \in D$ (in particular, $D = Z$ for a problem without constraints), A, A_h bounded linear operators acting from Z into U with $\|A - A_h\| \leq h$, $h \geq 0$. We are to construct an approximate solution of the equation

$$Az = u \tag{1.1}$$

belonging to D, given a set of data $\{A_h, u_\delta, \eta\}$, $\eta = (\delta, h)$, where $\delta > 0$ is the error in specifying the righthand side u_δ of (1.1), i.e. $\|u_\delta - \overline{u}\| \leq \delta$, $\overline{u} = A\overline{z}$. Here, $\overline{z}$ is the exact solution of (1.1), $\overline{z} \in D$, with $\overline{u}$ at the righthand side. We introduce the *smoothing functional* [165]:

$$M^\alpha[z] = \|A_h z - u_\delta\|^2 + \alpha \|z\|^2 \tag{1.2}$$

($\alpha > 0$ a *regularization parameter*) and consider the following extremal problem:
find

$$\inf_{z \in D} M^\alpha[z]. \tag{1.3}$$

LEMMA 1.1. *For any $\alpha > 0$, $u_\delta \in U$ and bounded linear operator A_h the problem (1.3) is solvable and has a unique solution $z_\eta^\alpha \in D$. Moreover,*

$$\|z_\eta^\alpha\| \leq \frac{\|u_\delta\|}{\sqrt{\alpha}}. \tag{1.4}$$

PROOF. Clearly, $M^\alpha[z]$ is twice Fréchet differentiable, and

$$(M^\alpha[z])' = 2\left(A_h^* A_h z + A_h^* u_\delta + \alpha z\right),$$
$$(M^\alpha[z])'' = 2\left(A_h^* A_h + \alpha E\right)$$

($A_h^* \colon U \to Z$ is the adjoint of A_h). For any $z \in Z$ we have $((M^\alpha[z])'' z, z) \geq 2\alpha \|z\|^2$, therefore $M^\alpha[z]$ is strongly convex. Consequently, on any (not necessarily bounded) closed set $D \subseteq Z$, $M^\alpha[z]$ attains its minimum at a unique point z_η^α [33].

Since $0 \in D$, we have

$$\inf_{z \in D} M^\alpha[z] \leq M^\alpha[0],$$

which implies (1.4). $\square$

So, $M^\alpha[z]$ is a strongly convex functional in a Hilbert space [27]. To find the extremal $z_\eta^\alpha \in D$ for $\alpha > 0$ fixed, it suffices to apply gradient methods for minimizing functionals with or without (if $D = Z$) constraints [33].

We recall that a necessary and sufficient condition for z_η^α to be a minimum point of $M^\alpha[z]$ on D is that [76]

$$\left((M^\alpha[z_\eta^\alpha])', z - z_\eta^\alpha\right) \geq 0, \qquad \forall z \in D.$$

If z_η^α is an interior point of D (or if $D = Z$), then this condition takes the form $(M^\alpha[z_\eta^\alpha])' = 0$, or

$$A_h^* A_h z_\eta^\alpha + \alpha z_\eta^\alpha = A_h^* u_\delta. \tag{1.5}$$

Thus, in this case we may solve the Euler equation (1.5) instead of minimizing $M^\alpha[z]$. The numerical aspects of realizing both approaches will be considered below.

2. Choice of the regularization parameter

The idea of constructing a regularizing algorithm using the extremal problem (1.3) for $M^\alpha[z]$ consists of constructing a function $\alpha = \alpha(\eta)$ such that $z_\eta^{\alpha(\eta)} \to \bar{z}$ as $\eta \to 0$, or, in other words, in matching the regularization parameter α with the error of specifying the initial data η.

THEOREM 1.1 ([165], [166], [169], [170]). *Let A be a bijective operator, $\bar{z} \in D$. Then $z_\eta^{\alpha(\eta)} \to \bar{z}$ as $\eta \to 0$, provided that $\alpha(\eta) \to 0$ in such a way that $(h+\delta)^2/\alpha(\eta) \to 0$.*

PROOF. Assume the contrary, i.e. $z_\eta^{\alpha(\eta)} \not\to \bar{z}$. This means that there are an $\epsilon > 0$ and a sequence $\eta_k \to 0$ such that $\|z_{\eta_k}^{\alpha(\eta_k)} - \bar{z}\| \geq \epsilon$. Since $\bar{z} \in D$, for any $\alpha > 0$,

$$M^\alpha[z_\eta^\alpha] = \inf_{z \in D} M^\alpha[z] \leq M^\alpha[\bar{z}] =$$
$$= \|A_h \bar{z} - u_\delta\|^2 + \alpha \|\bar{z}\|^2 =$$
$$= \|A_h \bar{z} - A\bar{z} + \bar{u} - u_\delta\|^2 + \alpha \|\bar{z}\|^2 \leq$$
$$\leq \left(h\|\bar{z}\| + \delta\right)^2 + \alpha \|\bar{z}\|^2.$$

Hence,

$$\|z_\eta^\alpha\|^2 \leq \frac{(h\|\overline{z}\| + \delta)^2}{\alpha} + \|\overline{z}\|^2. \tag{1.6}$$

According to the conditions of the theorem there is a constant C, independent of η for $\delta \leq \delta_0$, $h \leq h_0$ (with $\delta_0 > 0$, $h_0 > 0$ certain positive numbers) such that

$$\frac{(h\|\overline{z}\| + \delta)^2}{\alpha(\eta)} \leq C.$$

Further, using the weak compactness of a ball in a Hilbert space [92], we can extract from $\{z_{\eta_k}^{\alpha(\eta_k)}\}_k$ a subsequence that converges weakly to $z^* \in D$ (since D is weakly closed). Without loss of generality we will assume that $z_{\eta_k}^{\alpha(\delta_k)} \longrightarrow^{\mathrm{wk}} z^*$. Using the weak lower semicontinuity of the norm [123], the condition $(h + \delta)^2/\alpha(\eta) \to 0$ and inequality (1.6), we can easily obtain:

$$\|z^*\| \leq \liminf_{k \to \infty} \|z_{\eta_k}^{\alpha(\eta_k)}\| \leq \limsup_{k \to \infty} \|z_{\eta_k}^{\alpha(\eta_k)}\| \leq \|\overline{z}\|. \tag{1.7}$$

Consider now the inequality

$$\|Az_{\eta_k}^{\alpha(\eta_k)} - A\overline{z}\| \leq \|Az_{\eta_k}^{\alpha(\eta_k)} - A_{h_k}z_{\eta_k}^{\alpha(\eta_k)}\| + \|A_{h_k}z_{\eta_k}^{\alpha(\eta_k)} - u_{\delta_k}\| + \|u_{\delta_k} - \overline{u}\| \leq$$
$$\leq h_k\|z_{\eta_k}^{\alpha(\eta_k)}\| + \|A_{h_k}z_{\eta_k}^{\alpha(\eta_k)} - u_{\delta_k}\| + \delta_k \leq$$
$$\leq h_k\left(C + \|\overline{z}\|^2\right)^{1/2} + \left(\left(h_k\|\overline{z}\| + \delta_k\right)^2 + \alpha(\eta_k)\|\overline{z}\|^2\right)^{1/2} + \delta_k.$$

By limit transition as $k \to \infty$, using that $\alpha(\eta_k) \to 0$ and the weak lower semicontinuity of the functional $\|Az - A\overline{z}\|$, we obtain $\|Az^* - A\overline{z}\| = 0$, i.e. $z^* = \overline{z}$ (since A is bijective). Then (1.7) implies that $\lim_{k \to \infty} \|z_{\eta_k}^{\alpha(\eta_k)}\| = \|\overline{z}\|$. A Hilbert space has the H-property (weak convergence and norm convergence imply strong convergence [95]), so

$$\lim_{k \to \infty} z_{\eta_k}^{\alpha(\eta_k)} = \overline{z}.$$

This contradiction proves the theorem. $\square$

REMARK. Suppose that A is not bijective. Define a *normal solution* $\overline{z}$ of (1.1) on D with corresponding righthand side $\overline{u} = A\overline{z}$ to be a solution of the extremal problem

$$\|\overline{z}\|^2 = \inf \|z\|^2, \qquad z \in \{z : z \in D, Az = \overline{u}\}.$$

The set $\{z : z \in D, Az = \overline{u}\}$ is convex and closed, the functional $f(z) = \|z\|^2$ is strongly convex. Consequently, the normal solution $\overline{z} \in D$ exists and is unique [33].

If, in Theorem 1.1, we drop the requirement that A be bijective, then the assertion of this theorem remains valid if we take $\overline{z}$ to be the normal solution of (1.1). Thus, in this case $z_\eta^{\alpha(\eta)} \to \overline{z}$ as $\eta \to 0$, where $\overline{z}$ is the solution of (1.1) with minimal norm.

Theorem 1.1 indicates an order of decrease of $\alpha(\eta)$ that is sufficient for constructing a regularizing algorithm. It is clear that when processing actual experimental data, the error level η is fixed and known. Below we will consider an approach making it possible to use a fixed value of the error of specifying the initial data in order to construct stable approximate solutions of (1.1).

We define the *incompatibility measure* of (1.1) with approximate data on $D \subseteq Z$ as

$$\mu_\eta(u_\delta, A_h) = \inf_{z \in D} \|A_h z - u_\delta\|.$$

Clearly, $\mu_\eta(u_\delta, A_h) = 0$ if $u_\delta \in \overline{A_h D}$ (where the bar means norm closure in the corresponding space).

LEMMA 1.2. *If* $\|u_\delta - \overline{u}\| \leq \delta$, $\overline{u} = A\overline{z}$, $\overline{z} \in D$, $\|A - A_h\| \leq h$, *then* $\mu_\eta(u_\delta, A_h) \to 0$ *as* $\eta \to 0$.

PROOF. This follows from the fact that

$$\mu_\eta(u_\delta, A_h) = \inf_{z \in D} \|A_h z - u_\delta\| \leq \|A_h \overline{z} - u_\delta\| \leq \delta + h\|\overline{z}\| \to 0$$

as $\eta \to 0$. □

In the sequel we will assume that the incompatibility measure can be computed with error $\kappa > 0$, i.e. instead of $\mu_\eta(u_\delta, A_h)$ we only know a $\mu_\eta^\kappa(u_\delta, A_h)$ such that

$$\mu_\eta(u_\delta, A_h) \leq \mu_\eta^\kappa(u_\delta, A_h) \leq \mu_\eta(u_\delta, A_h) + \kappa.$$

We will assume that the error κ in determining the incompatibility measure matches the error of specifying the initial data η in the sense that $\kappa = \kappa(\eta) \to 0$ as $\eta \to 0$ (e.g., $\kappa(\eta) = h + \delta$).

We introduce the following function, called the *generalized discrepancy* [57], [59]:

$$\rho_\eta^\kappa(\alpha) = \|A_h z_\eta^\alpha - u_\delta\|^2 - \left(\delta + h\|z_\eta^\alpha\|\right)^2 - \left(\mu_\eta^\kappa(u_\delta, A_h)\right)^2.$$

We will now state the so-called *generalized discrepancy principle* for choosing the regularization parameter. Suppose the condition

$$\|u_\delta\|^2 > \delta^2 + \left(\mu_\eta^\kappa(u_\delta, A_h)\right)^2 \tag{1.8}$$

is not satisfied. Then we can take $z_\eta = 0$ as approximate solution of (1.1). If condition (1.8) is satisfied, then the generalized discrepancy has a positive zero, or a root $\alpha^* > 0$ (see §4), i.e.

$$\|A_h z_\eta^{\alpha^*} - u_\delta\|^2 = \left(\delta + h\|z_\eta^{\alpha^*}\|\right)^2 + \left(\mu_\eta^\kappa(u_\delta, A_h)\right)^2. \tag{1.9}$$

In this case we obtain an approximate solution $z_\eta = z_\eta^{\alpha^*}$ of (1.1); moreover, as will be shown below, $z_\eta = z_\eta^{\alpha^*}$ is uniquely defined.

THEOREM 1.2. *Let A be a bijective operator, $\eta = (\delta, h) \to 0$, such that $\|A - A_h\| \leq h$, $\|u_\delta - \overline{u}\| \leq \delta$, $\overline{u} = A\overline{z}$, $\overline{z} \in D$. Then $\lim_{\eta \to 0} z_\eta = \overline{z}$, where z_η is chosen in accordance with the generalized discrepancy principle.*

If A is not bijective (i.e. $\mathrm{Ker}\, A \neq \{0\}$), then the assertion above remains valid if $\overline{z}$ is taken to be the normal solution of (1.1).

PROOF. If $\overline{z} = 0$, then $\|u_\delta\| \leq \delta$ and condition (1.8) is not satisfied. In this case $z_\eta = 0$ and the theorem has been proved.

Suppose now that $\overline{z} \neq 0$. Then, since $\delta^2 + \left(\mu_\eta^\kappa(u_\delta, A_h)\right)^2 \to 0$ as $\eta \to 0$ (see Lemma 1.2 and the definitions of μ_η^κ and κ), $\|u_\delta\| \to \|\overline{u}\| \neq 0$, then condition (1.8) will be satisfied for sufficiently small η at least.

In this case the scheme of the proof of Theorem 1.2 does not differ from the scheme of proving Theorem 1.1. Again we assume that $z_\eta^{\alpha^*(\eta)} \not\to \overline{z}$. This means that there are an $\epsilon > 0$ and a subsequence $\eta_k \to 0$ such that $\|\overline{z} - z_{\eta_k}^{\alpha^*(\eta_k)}\| \geq \epsilon$. By the extremal properties, $z_{\eta_k}^{\alpha^*(\eta_k)} \in D$, we obtain $(\alpha^*(\eta_k) \equiv \alpha_k^*)$:

$$\|A_{h_k} z_{\eta_k}^{\alpha_k^*} - u_{\delta_k}\|^2 + \alpha_k^* \|z_{\eta_k}^{\alpha_k^*}\|^2 \leq \|A_{h_k}\overline{z} - u_{\delta_k}\|^2 + \alpha_k^* \|\overline{z}\|^2.$$

Using the generalized discrepancy principle (1.9) we obtain

$$\left(\delta_k + h_k \|z_{\eta_k}^{\alpha_k^*}\|\right)^2 + \alpha_k^* \|z_{\eta_k}^{\alpha_k^*}\|^2 \leq \left(\delta_k + h_k \|z_{\eta_k}^{\alpha_k^*}\|\right)^2 + \left(\mu_{\eta_k}^{\kappa_k}(u_{\delta_k}, A_{h_k})\right)^2 + \alpha_k^* \|z_{\eta_k}^{\alpha_k^*}\|^2 \leq$$

$$\leq \left(\delta_k + h_k \|\overline{z}\|\right)^2 + \alpha_k^* \|\overline{z}\|^2.$$

Thus,

$$f\left(\|z_{\eta_k}^{\alpha_k^*}\|\right) \leq f\left(\|\overline{z}\|\right),$$

where $f(x) = (A + Bx)^2 + Cx^2$, $A = \delta_k \geq 0$, $B = h_k \geq 0$, $C = \alpha_k^* > 0$. The function $f(x)$ is strictly monotone for $x > 0$, and hence

$$\|z_{\eta_k}^{\alpha_k^*}\| \leq \|\overline{z}\|.$$

Taking into account, as in the proof of Theorem 1.1, that $z_{\eta_k}^{\alpha_k^*} \overset{\text{wk}}{\longrightarrow} z^*$, $z^* \in D$, we arrive at

$$\|z^*\| \leq \liminf_{k \to \infty} \|z_{\eta_k}^{\alpha_k^*}\| \leq \limsup_{k \to \infty} \|z_{\eta_k}^{\alpha_k^*}\| \leq \|\overline{z}\|,$$

which is similar to (1.7).

To finish the proof it remains, as in the proof of Theorem 1.1, to show that $z^* = \overline{z}$. This follows from the inequality

$$\|Az_{\eta_k}^{\alpha_k^*} - A\overline{z}\| \leq h_k \|z_{\eta_k}^{\alpha_k^*}\| + \left(\left(\delta_k + h_k \|z_{\eta_k}^{\alpha_k^*}\|\right)^2 + \left(\mu_{\eta_k}^{\kappa_k}(u_\delta, A_{h_k})\right)^2\right)^{1/2} + \delta_k \leq$$

$$\leq h_k \|\overline{z}\| + \left(\left(\delta_k + h_k \|\overline{z}\|\right)^2 + \left(\mu_{\eta_k}^{\kappa_k}(u_\delta, A_{h_k})\right)^2\right)^{1/2} + \delta_k$$

and the fact that

$$h_k \|\overline{z}\| + \left(\left(\delta_k + h_k \|\overline{z}\|\right)^2 + \left(\mu_{\eta_k}^{\kappa_k}(u_\delta, A_{h_k})\right)^2\right)^{1/2} + \delta_k \to 0$$

as $\eta_k \to 0$. $\square$

In [162] it was noted that in the statement of the generalized discrepancy principle we can put $\mu_\eta^\kappa(u_\delta, A_h) = 0$ even if $u_\delta \notin A_h D$. In fact, we can define the following *generalized discrepancy*:

$$\rho_\eta(\alpha) = \|A_h z_\eta^\alpha - u_\delta\|^2 - \left(\delta + h\|z_\eta^\alpha\|\right)^2.$$

Condition (1.8) takes the form

$$\|u_\delta\| > \delta. \tag{1.10}$$

We can now state the *generalized discrepancy principle* as follows.

1. If the condition $\|u_\delta\| > \delta$ is not fulfilled, we take $z_\eta = 0$ as approximate solution of (1.1).
2. If the condition $\|u_\delta\| > \delta$ is fulfilled, then:
    a) if there is an $\alpha^* > 0$ which is a zero of the function $\rho_\eta(\alpha)$, then we take $z_\eta^{\alpha^*}$ as solution;
    b) if $\rho_\eta(\alpha) > 0$ for all $\alpha > 0$, then we take $z_\eta = \lim_{\alpha \downarrow 0} z_\eta^\alpha$ as approximate solution.

THEOREM 1.3. *If A is a bijective operator, then the above-mentioned algorithm is regularizing.*

If A is not bijective, then the regularized approximate solutions z_η converge to the normal solution of (1.1) on D.

PROOF. It is obvious (see Theorem 1.2) that the theorem holds for $\overline{z} = 0$. If $\overline{z} \neq 0$ and $\rho_\eta(\alpha)$ has a zero, the theorem has also been proved, since, proceeding as in the proof of Theorem 1.2, we can readily prove that

$$\left(\delta_k + h_k\|z_{\eta_k}^{\alpha_k^*}\|\right)^2 + \alpha_k^*\|z_{\eta_k}^{\alpha_k^*}\|^2 = \|A_{h_k} z_{\eta_k}^{\alpha_k^*} - u_{\delta_k}\|^2 + \alpha_k^*\|z_{\eta_k}^{\alpha_k^*}\|^2 \le$$
$$\le \|A_{h_k}\overline{z} - u_{\delta_k}\|^2 + \alpha_k^*\|\overline{z}\|^2 \le$$
$$\le \left(\delta_k + h_k\|\overline{z}\|\right)^2 + \alpha_k^*\|\overline{z}\|^2,$$

i.e. $\|z_\eta^{\alpha^*}\| \le \|\overline{z}\|$. The remainder of the proof of Theorem 1.2 is practically unchanged. So, it remains to consider the case when $\rho_\eta(\alpha) > 0$ for all $\alpha > 0$. Since

$$\|A_h z_\eta^\alpha - u_\delta\|^2 + \alpha\|z_\eta^\alpha\| \le \|A_h \overline{z} - u_\delta\|^2 + \alpha\|\overline{z}\|^2,$$
$$\|A_h z_\eta^\alpha - u_\delta\| > \delta + h\|z_\eta^\alpha\|,$$
$$\|A_h \overline{z} - u_\delta\| \le \delta + h\|\overline{z}\|,$$

we have

$$\left(\delta + h\|z_\eta^\alpha\|\right)^2 + \alpha\|z_\eta^\alpha\|^2 \le \left(\delta + h\|\overline{z}\|\right)^2 + \alpha\|\overline{z}\|^2.$$

As before, this implies that $\|z_\eta^\alpha\| \le \|\overline{z}\|$ for all $\alpha > 0$. So, from any sequence α_n converging to zero we can extract a subsequence α_n' such that $z_\eta^{\alpha_n'}$ converges weakly to $z_\eta \in D$ (since D is weakly closed). Without loss of generality we will assume that $z_\eta^{\alpha_n} \xrightarrow{\text{wk}} z_\eta$. In §4 we will show that

$$\lim_{\alpha \downarrow 0} M^\alpha[z_\eta^\alpha] = \left(\mu_\eta(u_\delta, A_h)\right)^2.$$

Therefore, since

$$\left(\mu_\eta(u_\delta, A_h)\right)^2 \leq \|A_h z_\eta^{\alpha_n} - u_\delta\|^2 + \alpha_n \|z_\eta^{\alpha_n}\|^2 \to \left(\mu_\eta^\kappa(u_\delta, A_h)\right)^2$$

as $\alpha_n \to 0$, we have

$$\|A_h z_\eta - u_\delta\| = \inf_{z \in D} \|A_h z - u_\delta\| = \mu_\eta(u_\delta, A_h)$$

and z_η is an extremal of $M^\alpha[z]$ for $\alpha = 0$ (i.e. z_η is a quasisolution of equation (1.1) with approximate data) satisfying the inequality $\|z_\eta\| \leq \|\bar{z}\|$, while $z_\eta^{\alpha_n}$ is a minimizing sequence for the functional $\|A_h z - u_\delta\|^2$ on D. The function $\|z_\eta^\alpha\|$ is monotonically nondecreasing in α (see §4) and bounded above by $\|\bar{z}\|$. Hence

$$\lim_{\alpha \downarrow 0} \|z_\eta^\alpha\| = a$$

exists and, moreover,

$$\lim_{n \to \infty} \|z_\eta^{\alpha_n}\| = a \geq \|z_\eta\|.$$

We will show that $\|z_\eta\| = a$. Assume the contrary. Then, from some N onwards, $\|z_\eta^{\alpha_n}\| > \|z_\eta\|$. But, since

$$\|A_h z_\eta^{\alpha_n} - u_\delta\|^2 + \alpha_n \|z_\eta^{\alpha_n}\|^2 \leq \left(\mu_\eta(u_\delta, A_h)\right)^2 + \alpha_n \|z_\eta\|^2$$

($z_\eta^{\alpha_n}$ is an extremal of $M^{\alpha_n}[z]$), we have $\|A_h z_\eta^{\alpha_n} - u_\delta\| \leq \mu_\eta(u_\delta, A_h)$, from some N onwards. By the definition of incompatibility measure,

$$\|A_h z_\eta^{\alpha_n} - u_\delta\| = \mu_\eta(u_\delta, A_h).$$

Therefore z_η is an extremal of $M^\alpha[z]$ for all α_n, from some N onwards (but defined in a unique manner), i.e.

$$\|z_\eta\| = \|z_\eta^{\alpha_n}\|.$$

This contradiction shows that $\|z_\eta\| = \lim_{n \to \infty} \|z_\eta^{\alpha_n}\|$, and thus $z_\eta^{\alpha_n}$ converges to z_η not only in the weak sense, but also in the strong sense. To prove the existence of $\lim_{\alpha \downarrow 0} z_\eta^\alpha$ it now suffices to prove that the limit z_η of the sequence $\{z_\eta^{\alpha_n}\}$ does not depend on the choice of the sequence $\{\alpha_n\}$. In fact, the limit of $\{z_\eta^{\alpha_n}\}$ is the extremal of $M^0[z]$ with minimal norm. Let $\bar{\bar{z}}_\eta$ be the extremal of $M^0[z]$ with minimal norm, i.e.

$$\|A_h \bar{\bar{z}}_\eta - u_\delta\|^2 = \left(\mu_\eta(u_\delta, A_h)\right)^2 \leq \|A_h z_\eta^{\alpha_n} - u_\delta\|^2.$$

It is clear that

$$\|A_h z_\eta^{\alpha_n} - u_\delta\|^2 + \alpha_n \|z_\eta^{\alpha_n}\|^2 \leq \|A_h \bar{\bar{z}}_\eta - u_\delta\|^2 + \alpha_n \|\bar{\bar{z}}_\eta\|^2.$$

Consequently, $\|z_\eta^{\alpha_n}\| \leq \|\bar{\bar{z}}_\eta\|$, and so

$$\|z_\eta\| = \|\lim_{n \to \infty} z_\eta^{\alpha_n}\| = \lim_{n \to \infty} \|z_\eta^{\alpha_n}\| \leq \|\bar{\bar{z}}_\eta\|.$$

Since z_η is an extremal of $M^0[z]$, we have $z_\eta = \bar{\bar{z}}_\eta$.

Thus, we have proved that $z_\eta = \lim_{\alpha \downarrow 0} z_\eta^\alpha$ is the solution of the problem
find

$$\inf \|z\|, \qquad z \in D, \quad \|A_h z - u_\delta\| = \mu_\eta(u_\delta, A_h).$$

The solution of this problem exists and is unique. This proves that the limit $\lim_{\alpha \downarrow 0} z_\eta^\alpha = z_\eta$ exists (where z_η is the normal pseudosolution of the equation $A_h z = u_\delta$). To finish the proof it suffices to note that

$$\|Az_\eta - \overline{u}\| \le \|Az_\eta - A_h z_\eta - u_\delta + u_\delta - \overline{u}\| \le$$
$$\le h\|z_\eta\| + \mu_\eta(u_\delta, A_h) + \delta \to 0$$

as $\eta \to 0$. $\square$

Note that in distinction to the algorithm based on Theorem 1.2, the given modification of the generalized discrepancy principle does not require one to compute $\mu_\eta^\kappa(u_\delta, A_h)$. Instead, the algorithm of Theorem 1.3 requires that equation (1.1) with exact data be solvable on D.

REMARKS. 1. In [90], [163] the optimality, with respect to order, of the generalized discrepancy principle on a compactum that is the image of a ball in a reflexive space under a compact map has been shown.

2. The above-proposed approach to solving linear problems with an approximately given operator can be regarded as a generalized least-squares method of Legendre [217] and Gauss [210], which is known to be unstable under perturbations of the operator (the matrix of the system of linear algebraic equations).

3. For systems of linear algebraic equations and linear equations in a Hilbert space, in [174]–[177] a method has been proposed making it possible not only to obtain an approximate solution of a problem with perturbed operator (matrix), but also to find the operator (matrix) realizing the given solution.

Stable methods for solving problems with approximately given operator were proposed in [169], [170]. The generalized discrepancy principle is an extension of the *discrepancy principle* for choosing the regularization parameter in accordance with the equality $\|Az_\alpha - u_\delta\| = \delta$ (or $\|Az_\alpha - u_\delta\| = C\delta$, $C > 1$), $\eta = (\delta, 0)$, $\delta > 0$, $\|u_\delta\| > \delta$ (see [10], [38], [88], [128], [131], [135]).

The generalized discrepancy principle for Hilbert spaces has been proposed and substantiated in [57], [59], [63], and for reflexive spaces in [203], [205]. The generalized discrepancy principle has been considered in [90], [100], [101], [122], [138], [163], [206]. Applications of the generalized discrepancy principle to the solution of practical problems, as well as model calculations, can be found in [71], [73], [94], [96], [118], [185]. The generalized discrepancy principle for nonlinear problems has been considered in [7], [112], [114]–[117].

In conclusion we consider the problem of stability of the solution of the extremal problem (1.3) under small perturbations of u_δ, A_h, α. The similar problem for an exactly given operator using the scheme of compact imbedding has been considered in [90]. Let $P(u_\delta, A_h, \alpha) = z_\eta^\alpha$ be the map from the product space $U \times \text{Hom}(Z, U) \times \mathbf{R}^+$ into the set $D \subseteq Z$ describing the solution of the problem of minimizing $M^\alpha[z]$ on D. Here, $\text{Hom}(Z, U)$ is the space of bounded linear operators from Z into U, equipped with the uniform operator norm, and $\mathbf{R}^+$ is the space of positive real numbers with the natural metric.

THEOREM 1.4. *The map P from $U \times \text{Hom}(Z, U) \times \mathbf{R}^+$ into Z is continuous.*

PROOF. If P would be not continuous, there would be a sequence (u_n, A_n, α_n) such that $(u_n, A_n, \alpha_n) \to (u, A, \alpha)$ but $z_n = \overset{\backprime}{P}(u_n, A_n, \alpha_n) \not\to z^\alpha = P(u, A, \alpha)$.

Without loss of generality we may assume that

$$\|z_n - z^\alpha\| \geq d > 0, \qquad d = \text{const}.$$

Since $z_n = P(u_n, A_n, \alpha_n)$, we have

$$\|A_n z_n - u_n\|^2 + \alpha_n \|z_n\|^2 = \inf_{z \in D} \left(\|A_n z - u_n\|^2 + \alpha_n \|z\|^2 \right).$$

By (1.4),

$$\|z_n\| \leq \frac{\sup_n \|u_n\|}{\inf_n \sqrt{\alpha_n}}.$$

Since $\alpha = \lim_{n \to \infty} \alpha_n \neq 0$, we can extract a subsequence from z_n that converges weakly to some $z^* \in D$. Without loss of generality we may assume that for the whole sequence z_n we have $z_n \overset{wk}{\longrightarrow} z^* \in D$ (since D is convex and closed and, hence, weakly closed) and $\|z_n\| \to a$. It can be readily seen that $A_n z_n \overset{wk}{\longrightarrow} A z^*$. In fact,

$$\|(A_n - A)z_n\| \leq \|A_n - A\| \frac{\sup_n \|u_n\|}{\inf_n \sqrt{\alpha_n}}.$$

Therefore $(A_n - A)z_n \overset{str}{\longrightarrow} 0$ as $n \to \infty$, and hence $(A_n - A)z_n \overset{wk}{\longrightarrow} 0$. Further, $A z_n \overset{wk}{\longrightarrow} A z^*$. Since A is a bounded linear operator, $A_n z_n \overset{wk}{\longrightarrow} A z^*$.

So,

$$
\begin{aligned}
\|A z^* - u\|^2 + \alpha |z^*\|^2 &\leq \liminf_{n \to \infty} \left(\|A_n z_n - u_n\|^2 + \alpha_n \|z_n\|^2 \right) \leq \\
&\leq \liminf_{n \to \infty} \left(\|A_n z^\alpha - u_n\|^2 + \alpha_n \|z^\alpha\|^2 \right) \leq \\
&\leq \lim_{n \to \infty} \left(\left(\|A - A_n\| \|z^\alpha\| + \|A_n z^\alpha - u_n\| \right)^2 + \alpha_n \|z^\alpha\|^2 \right) = \\
&= \|A z^\alpha - u\|^2 + \alpha \|z^\alpha\|^2.
\end{aligned}
\tag{1.11}
$$

Since $z^* \in D$, we have $z^* = z^\alpha$ and all inequalities in the above chain of inequalities become equalities. By limit transition as $n \to \infty$ in the inequality

$$\|A_n z_n - u_n\|^2 + \alpha_n \|z_n\|^2 \leq \|A_n z^\alpha - u_n\|^2 + \alpha_n \|z^\alpha\|^2$$

and taking into account that $\|z_n\| \to a$, we obtain

$$\|A z^* - u\|^2 + \alpha a^2 = \|A z^\alpha - u\|^2 + \alpha \|z^\alpha\|^2.$$

Comparing this equality with (1.11), we obtain $\|z^*\| = a = \|z^\alpha\|$, i.e. $z_n \overset{wk}{\longrightarrow} z^\alpha$ and $\|z_n\| \to \|z^\alpha\|$, in other words, $z_n \overset{str}{\longrightarrow} z^\alpha$, contradicting the assumption. $\square$

3. Equivalence of the generalized discrepancy principle and the generalized discrepancy method

To better understand the meaning of choosing the regularization parameter in accordance with the generalized discrepancy principle, we show the equivalence of the latter with the *generalized discrepancy method*, i.e. with the following extremal problem with constraints:

find

$$\inf \|z\|, \qquad z \in \left\{ z : z \in D, \ \|A_h z - u_\delta\|^2 \leq (\delta + h\|z\|)^2 + \left(\mu_\eta^\kappa(u_\delta, A_h)\right)^2 \right\}. \tag{1.12}$$

The generalized discrepancy method was introduced for the first time in [58] to solve nonlinear ill-posed problems with approximately given operator on a concrete function space. It is a generalization of the discrepancy method ($h = 0$) for solving similar problems with exactly given operator. The idea of the discrepancy method was first expressed in [218]; however, a strict problem statement and a substantiation of the method were first given in [88]. The further development of the circle of ideas related to applying the discrepancy method to the solution of ill-posed problems was given in [34], [36], [37], [89], [121], [139], [161], [209]. In [102] and [204] the generalized discrepancy method in reflexive spaces has been studied.

The equivalence of the generalized discrepancy method and the generalized discrepancy principle was proved first in [162], although under certain superfluous restrictions. Below we will adhere to the scheme of proof given in [205].

THEOREM 1.5. *Let A, A_h be bounded linear operators from Z into U, let D be a closed convex set containing the point 0, $D \subseteq Z$, let $\|A - A_h\| \leq h$, $\|u_\delta - \overline{u}\| \leq \delta$, $\overline{u} = A\overline{z}$, $\overline{z} \in D$.*

Then the generalized discrepancy principle and the generalized discrepancy method are equivalent, i.e. the solution of (1.1) chosen under the conditions (1.8)–(1.9) and the solution of the extremal problem (1.12) coincide.

PROOF. Suppose (1.8) is not satisfied. Then

$$0 \in \left\{ z : z \in D, \ \|A_h z - u_\delta\|^2 \leq \left(\delta + h\|z\|\right)^2 + \left(\mu_\eta^\kappa(u_\delta, A_h)\right)^2 \right\},$$

and the solution of (1.12) is $\hat{z}_\eta = 0$. Since in this case the generalized discrepancy principle also leads to $z_\eta = 0$, it remains to prove their equivalence in case (1.8) is satisfied.

In that case (1.12) is equivalent to the following problem:
find

$$\inf \|z\|, \qquad z \in \left\{ z : z \in D, \ \|A_h z - u_\delta\|^2 = \left(\delta + h\|z\|\right)^2 + \left(\mu_\eta^\kappa(u_\delta, A_h)\right)^2 \right\}.$$

Indeed, $0 \notin \{z : z \in D, \ \|A_h z - u_\delta\|^2 \leq (\delta + h\|z\|)^2 + \left(\mu_\eta^\kappa(u_\delta, A_h)\right)^2\}$. Let $\hat{z}_\eta \in D$ be a solution of (1.12) such that

$$\|A_h \hat{z}_\eta - u_\delta\|^2 < \left(\delta + h\|\hat{z}_\eta\|\right)^2 + \left(\mu_\eta^\kappa(u_\delta, A_h)\right)^2.$$

The function

$$f(x) = \|A_h z_x - u_\delta\|^2 - \left(\delta + h\|z_x\|\right)^2 - \left(\mu_\eta^\kappa(u_\delta, A_h)\right)^2$$

of the real variable x, where $z_x = x\hat{z}_\eta$, is continuous on $[0, 1]$, and $f(0) > 0$, $f(1) < 0$. Thus, there is an $\overline{x} \in (0, 1)$ such that $f(\overline{x}) = 0$, i.e.

$$\|A_h z_{\overline{x}} - u_\delta\|^2 = \left(\delta + h\|z_{\overline{x}}\|\right)^2 + \left(\mu_\eta^\kappa(u_\delta, A_h)\right)^2.$$

But

$$\|z_{\overline{x}}\| = \overline{x}\|\hat{z}_\eta\| < \|\hat{z}_\eta\|,$$

which contradicts the fact that $\hat{z}_\eta$ is the solution of (1.12).

We now turn to the generalized discrepancy principle. Let $z_\eta = z_\eta^{\alpha^*}$, $\alpha^* > 0$, where $z_\eta^{\alpha^*}$ is uniquely defined and $z_\eta^{\alpha^*} \neq 0$ and (as we will show below)

$$\|u_\delta\|^2 > \left(\delta + h\|z_\eta^{\alpha^*}\|\right)^2 + \left(\mu_\eta^\kappa(u_\delta, A_h)\right)^2.$$

Instead of $M^\alpha[z]$ we consider the functional

$$\tilde{M}^\lambda[z] = \lambda \left[\|A_h z - u_\delta\|^2 - \left(\delta + h\|z_\eta^{\alpha^*}\|\right)^2 - \left(\mu_\eta^\kappa(u_\delta, A_h)\right)^2\right] + \|z\|^2,$$

where $\lambda = 1/\alpha$. By (1.9),

$$\tilde{M}^{\lambda^*}[z_\eta^{\alpha^*}] = \tilde{M}^\lambda[z_\eta^{\alpha^*}] = \|z_\eta^{\alpha^*}\|^2$$

for all $\lambda > 0$ (here, $\lambda^* = 1/\alpha^*$).

Since $z_\eta^{\alpha^*}$ is an extremal of $M^{\alpha^*}[z]$, and hence of $M^{\lambda^*}[z]$, for all $z \in D$,

$$\tilde{M}^{\lambda^*}[z_\eta^{\alpha^*}] \leq \tilde{M}^{\lambda^*}[z].$$

Hence $(z_\eta^{\alpha^*}, \lambda^*)$ is a saddle point of $\tilde{M}^\lambda[z]$. By the Kuhn-Tucker theorem [90], $z_\eta^{\alpha^*}$ is a solution of the problem

find

$$\inf \|z\|, \qquad z \in \left\{z : z \in D, \|A_h z - u_\delta\|^2 = \left(\delta + h\|z_\eta^{\alpha^*}\|\right)^2 + \left(\mu_\eta^\kappa(u_\delta, A_h)\right)^2\right\}.$$

Indeed, suppose the set of constraints contains a z' such that

$$z' \in \left\{z : z \in D, \|A_h z - u_\delta\|^2 = \left(\delta + h\|z_\eta^{\alpha^*}\|\right)^2 + \left(\mu_\eta^\kappa(u_\delta, A_h)\right)^2\right\}$$

and

$$\|z'\| < \|z_\eta^{\alpha^*}\|.$$

Then

$$\tilde{M}^{\lambda^*}[z'] = \|z'\|^2 < \tilde{M}^{\lambda^*}[z_\eta^{\alpha^*}] = \|z_\eta^{\alpha^*}\|^2,$$

contradicting the inequality $\tilde{M}^{\lambda^*}[z_\eta^{\alpha^*}] \leq \tilde{M}^{\lambda^*}[z]$, which holds for all $z \in D$.

By the condition

$$0 \notin \left\{z : z \in D, \|A_h z - u_\delta\|^2 \leq \left(\delta + h\|z_\eta^{\alpha^*}\|\right)^2 + \left(\mu_\eta^\kappa(u_\delta, A_h)\right)^2\right\},$$

which follows from (1.8)–(1.9) and the monotonicity of $\beta_\eta(\alpha)$ and limit transition as $\alpha \to \infty$ (proven below, in Lemma 1.3), the extremal problem above is equivalent to the extremal problem (the proof of this equivalence is similar to that of the equivalence between the two extremal problems mentioned at the beginning of this proof):
find

$$\inf \|z\|, \qquad z \in \left\{ z : z \in D, \|A_h z - u_\delta\|^2 \leq \left(\delta + h\|z_\eta^{\alpha^*}\| \right)^2 + \left(\mu_\eta^\kappa(u_\delta, A_h) \right)^2 \right\}. \tag{1.13}$$

The solution of this problem exists and is unique because the set $\{ z : z \in D,$ $\|A_h z - u_\delta\|^2 \leq \left(\delta + h\|z_\eta^{\alpha^*}\| \right)^2 + \left(\mu_\eta^\kappa(u_\delta, A_h) \right)^2 \}$ is closed and convex in the Hilbert space Z. Hence its solution is $z_\eta^{\alpha^*}$.

We show that $z_\eta^{\alpha^*}$ is the solution of (1.12). Assume that the set of constraints of (1.12) contains an element $\hat{z}$ such that $\|\hat{z}\| < \|z^{\alpha^*}\|$ (note that, by (1.9), $z_\eta^{\alpha^*}$ satisfies the constraints of (1.12)). Then $\hat{z} \in D$ and

$$\|A_h \hat{z} - u_\delta\| \leq \left(\delta + h\|\hat{z}\| \right)^2 + \left(\mu_\eta^\kappa(u_\delta, A_h) \right)^2.$$

But $z_\eta^{\alpha^*}$ is the solution of (1.13), and

$$\left\{ z : z \in D, \|A_h z - u_\delta\|^2 \leq \left(\delta + h\|\hat{z}\| \right)^2 + \left(\mu_\eta^\kappa(u_\delta, A_h) \right)^2 \right\} \subseteq$$
$$\subseteq \left\{ z : z \in D, \|A_h z - u_\delta\|^2 \leq (\delta + h\|z_\eta^{\alpha^*}\|)^2 + \left(\mu_\eta^\kappa(u_\delta, A_h) \right)^2 \right\}.$$

Thus, $\|\hat{z}\| \geq \|z_\eta^{\alpha^*}\|$, since $\|\hat{z}\| \geq \|\hat{\hat{z}}\|$, where $\hat{\hat{z}}$ is the solution of the problem
find

$$\inf \|z\|, \qquad z \in \left\{ z : z \in D, \|A_h z - u_\delta\|^2 \leq \left(\delta + h\|\hat{z}\| \right)^2 + \left(\mu_\eta^\kappa(u_\delta, A_h) \right)^2 \right\}.$$

The contradiction obtained shows that $z_\eta^{\alpha^*}$ is the unique solution of (1.12). $\square$

So, the solution of the extremal problem (1.12) with nonconvex constraints can be reduced to a convex programming problem: minimize the functional $M^\alpha[z]$ by choosing the regularization parameter in accordance with the generalized discrepancy principle.

REMARKS. 1. Let $D = Z$, $\overline{A_h Z} = U$. Then in the statement of the generalized discrepancy principle and in (1.12) we may put $\mu_\eta^\kappa(u_\delta, A_h) = 0$.

2. We can readily show (by simple examples) that the modification of the generalized discrepancy principle considered in Theorem 1.3 is, in general, not equivalent to the problem (1.12) with μ_η^κ absent, i.e. it is not equivalent to the problem:
find

$$\inf \|z\|, \qquad z \in \left\{ z : z \in D, \|A_h z - u_\delta\| \leq \delta + h\|z\| \right\}.$$

The relation between the generalized discrepancy method and the generalized discrepancy principle for nonlinear ill-posed problems has been investigated in [116].

4. The generalized discrepancy and its properties

In this Section we study in detail the properties of certain auxiliary functions of the regularization parameter $\alpha > 0$:

$$\Phi_\eta(\alpha) = M^\alpha[z_\eta^\alpha] = \|A_h z_\eta^\alpha - u_\delta\|^2 + \alpha\|z_\eta^\alpha\|^2, \tag{1.14}$$

$$\gamma_\eta(\alpha) = \|z_\eta^\alpha\|^2, \tag{1.15}$$

$$\beta_\eta(\alpha) = \|A_h z_\eta^\alpha - u_\delta\|^2, \tag{1.16}$$

as well as properties of the generalized discrepancy $\rho_\eta^\kappa(\alpha)$ introduced in §2.

LEMMA 1.3. *The functions $\Phi_\eta(\alpha)$, $\gamma_\eta(\alpha)$, $\beta_\eta(\alpha)$ have, as functions of the parameter α, the following properties*

1. *They are continuous for $\alpha > 0$.*
2. *$\Phi_\eta(\alpha)$ is concave, differentiable, and $\Phi'_\eta(\alpha) = \gamma_\eta(\alpha)$.*
3. *$\gamma_\eta(\alpha)$ is monotonically nonincreasing and $\Phi_\eta(\alpha)$, $\beta_\eta(\alpha)$ are monotonically nondecreasing for $\alpha > 0$. Moreover, in the interval $(0, \alpha_0)$ in which $z_\eta^{\alpha_0} \neq 0$, the function $\Phi_\eta(\alpha)$ is strictly monotone.*
4. *The following limit relations hold:*

$$\lim_{\alpha \to \infty} \gamma_\eta(\alpha) = \lim_{\alpha \to \infty} \alpha\gamma_\eta(\alpha) = 0,$$

$$\lim_{\alpha \to \infty} \Phi_\eta(\alpha) = \lim_{\alpha \to \infty} \beta_\eta(\alpha) = \|u_\delta\|^2,$$

$$\lim_{\alpha \downarrow 0} \alpha\gamma_\eta(\alpha) = 0,$$

$$\lim_{\alpha \downarrow 0} \Phi_\eta(\alpha) = \lim_{\alpha \downarrow 0} \beta_\eta(\alpha) = \big(\mu_\eta(u_\delta, A_h)\big)^2.$$

PROOF. Assertions 2 and 3 readily follow from (1.18), which was first obtained and applied to the study of the auxiliary functions in [127]–[131]. Fixing $\alpha' \in (0, \alpha)$ and subtracting from the obvious inequality

$$\Phi_\eta(\alpha) = \|A_h z_\eta^\alpha - u_\delta\|^2 + \alpha\|z_\eta^\alpha\|^2 \leq \|A_h z_\eta^{\alpha'} - u_\delta\|^2 + \alpha\|z_\eta^{\alpha'}\|^2$$

the similar expression for $\Phi_\eta(\alpha')$, we obtain

$$\Phi_\eta(\alpha) - \Phi_\eta(\alpha') \leq (\alpha - \alpha')\|z_\eta^{\alpha'}\|^2. \tag{1.17}$$

We similarly obtain the inequality

$$\Phi_\eta(\alpha') - \Phi_\eta(\alpha) \leq (\alpha' - \alpha)\|z_\eta^\alpha\|^2.$$

Together they imply

$$\|z_\eta^\alpha\|^2 \leq \frac{\Phi_\eta(\alpha) - \Phi_\eta(\alpha')}{\alpha - \alpha'} \leq \|z_\eta^{\alpha'}\|^2. \tag{1.18}$$

Since $\|z_\eta^\alpha\|^2 \geq 0$, we see that $\Phi_\eta(\alpha)$ is monotonically nondecreasing, while if $\|z_\eta^\alpha\| \neq 0$ on the interval $(0, \alpha_0)$, then $\Phi_\eta(\alpha)$ increases monotonically on this interval. Equation (1.18) implies also that if $\alpha' \in (0, \alpha)$, then $\|z_\eta^\alpha\|^2 \leq \|z_\eta^{\alpha'}\|^2$, which means that $\gamma_\eta(\alpha)$ does not increase.

Inequality (1.4) implies that if we fix α_0, $0 < \alpha_0 < \alpha' < \alpha$, then

$$\|z_\eta^{\alpha'}\| \le \frac{\|u_\delta\|}{\sqrt{\alpha_0}}.$$

This and (1.17) imply the continuity of $\Phi_\eta(\alpha)$ for any $\alpha > 0$.

Further, (1.4) implies that $\lim_{\alpha \to \infty} \gamma_\eta(\alpha) = 0$. By the continuity of A_h,

$$\lim_{\alpha \to \infty} \beta_\eta(\alpha) = \lim_{\alpha \to \infty} \|A_h z_\eta^\alpha - u_\delta\|^2 = \|u_\delta\|^2.$$

Since $M^\alpha[z_\eta^\alpha] \le M^\alpha[0] = \|u_\delta\|^2$, we have

$$\lim_{\alpha \to \infty} \alpha \gamma_\eta(\alpha) = 0, \qquad \lim_{\alpha \to \infty} \Phi_\eta(\alpha) = \|u_\delta\|^2.$$

We now show that the discrepancy $\beta_\eta(\alpha)$ is monotonically nondecreasing. To this end it suffices to note that for $\alpha' \in (0, \alpha)$,

$$\|A_h z_\eta^{\alpha'} - u_\delta\|^2 + \alpha'\|z_\eta^{\alpha'}\|^2 \le \|A_h z_\eta^\alpha - u_\delta\|^2 + \alpha'\|z_\eta^\alpha\|^2 \le$$
$$\le \|A_h z_\eta^\alpha - u_\delta\|^2 + \alpha'\|z_\eta^{\alpha'}\|^2.$$

The second of these inequalities is a consequence of the nonincrease of $\gamma_\eta(\alpha)$. To prove the continuity of $\gamma_\eta(\alpha)$ we use the condition

$$\left((M^\alpha[z_\eta^\alpha])', z - z_\eta^\alpha\right) \ge 0, \qquad \forall z \in D, \quad \forall \alpha > 0,$$

or

$$\left(A_h^* A_h z_\eta^\alpha + \alpha z_\eta^\alpha - A_h^* u_\delta, z - z_\eta^\alpha\right) \ge 0,$$
$$\left(A_h^* A_h z_\eta^{\alpha'} + \alpha' z_\eta^{\alpha'} - A_h^* u_\delta, z - z_\eta^{\alpha'}\right) \ge 0, \qquad \forall z \in D.$$

Putting $z = z_\eta^{\alpha'}$ in the first of these and $z = z_\eta^\alpha$ in the second and adding the obtained inequalities gives

$$\left(A_h^* A_h(z_\eta^\alpha - z_\eta^{\alpha'}) + \alpha z_\eta^\alpha - \alpha' z_\eta^{\alpha'}, z_\eta^{\alpha'} - z_\eta^\alpha\right) \ge 0,$$

or

$$\left(A_h^* A_h(z_\eta^\alpha - z_\eta^{\alpha'}) + \alpha z_\eta^\alpha - \alpha z_\eta^{\alpha'} + \alpha z_\eta^{\alpha'} - \alpha' z_\eta^{\alpha'}, z_\eta^{\alpha'} - z_\eta^\alpha\right) \le 0,$$

or

$$\left|A_h(z_\eta^\alpha - z_\eta^{\alpha'})\right|^2 + \alpha \left|z_\eta^\alpha - z_\eta^{\alpha'}\right|^2 \le (\alpha' - \alpha)(z_\eta^{\alpha'}, z_\eta^{\alpha'} - z_\eta^\alpha).$$

Hence,

$$\|z_\eta^{\alpha'}\| - \|z_\eta^\alpha\| \le \|z_\eta^\alpha - z_\eta^{\alpha'}\| \le$$
$$\le \frac{1}{\sqrt{\alpha}} |\alpha' - \alpha| \left\{\|z_\eta^{\alpha'}\| \left(\|z_\eta^\alpha\| + \|z_\eta^{\alpha'}\|\right)\right\}^{1/2} \le$$
$$\le \frac{\sqrt{2}}{\alpha_0^{3/2}} \|u_\delta\| |\alpha' - \alpha|,$$

i.e. $\gamma_\eta(\alpha) = \|z_\eta^\alpha\|^2$ is continuous (even Lipschitz continuous). The continuity of $\beta_\eta(\alpha)$ follows from the continuity of $\Phi_\eta(\alpha)$, $\gamma_\eta(\alpha)$.

Inequality (1.8) and the continuity of $\gamma_\eta(\alpha)$ imply that $\Phi_\eta(\alpha)$ is differentiable for all $\alpha > 0$, and $\Phi'_\eta(\alpha) = \gamma_\eta(\alpha)$. So, $\Phi_\eta(\alpha)$ is concave because its derivative $\gamma_\eta(\alpha)$ is monotonically nonincreasing [91].

Since $\lim_{\alpha \to \infty} \|z_\eta^\alpha\| = 0$ and $\|z_\eta^\alpha\|$ does not grow as α increases, it is easily seen that if $z_\eta^{\alpha_0} = 0$ for a certain $\alpha_0 \in (0, \infty)$, then $z_\eta^\alpha = 0$ for all $\alpha > \alpha_0$. If $z_\eta^{\alpha_0} \neq 0$ for a certain $\alpha_0 > 0$, then $z_\eta^\alpha \neq 0$ for all $\alpha \in (0, \alpha_0)$.

We will now consider the behavior of the functions $\Phi_\eta(\alpha)$, $\gamma_\eta(\alpha)$, $\beta_\eta(\alpha)$ as $\alpha \downarrow 0$. Take $\epsilon > 0$ arbitrary. Then we can find a $z^\epsilon \in D$ such that

$$\mu_\eta(u_\delta, A_h) \leq \|A_h z^\epsilon - u_\delta\| \leq \mu_\eta(u_\delta, A_h) + \epsilon.$$

The inequality

$$\beta_\eta(\alpha) \leq \Phi_\eta(\alpha) \leq \|A_h z^\epsilon - u_\delta\|^2 + \alpha \|z^\epsilon\|^2$$

is obvious, and implies that

$$\limsup_{\alpha \downarrow 0} \beta_\eta(\alpha) \leq \lim_{\alpha \downarrow 0} \Phi_\eta(\alpha) \leq \left(\mu_\eta(u_\delta, A_h) + \epsilon \right)^2.$$

However, since

$$\beta_\eta(\alpha) \geq \inf_{z \in D} \|A_h z - u_\delta\|^2 = \left(\mu_\eta(u_\delta, A_h) \right)^2,$$

we have

$$\liminf_{\alpha \downarrow 0} \beta_\eta(\alpha) \geq \left(\mu_\eta(u_\delta, A_h) \right)^2.$$

Since $\epsilon > 0$ is arbitrarily chosen, we obtain

$$\lim_{\alpha \downarrow 0} \beta_\eta(\alpha) = \left(\mu_\eta(u_\delta, A_h) \right)^2,$$

and also

$$\lim_{\alpha \downarrow 0} \alpha \|z_\eta^\alpha\|^2 = 0, \qquad \lim_{\alpha \downarrow 0} \Phi_\eta(\alpha) = \left(\mu_\eta(u_\delta, A_h) \right)^2. \qquad \square$$

COROLLARY. *The generalized discrepancy*

$$\rho_\eta^\kappa(\alpha) = \beta_\eta(\alpha) - \left(\delta + h\sqrt{\gamma_\eta(\alpha)} \right)^2 - \left(\mu_\eta^\kappa(u_\delta, A_h) \right)^2$$

has the following properties:

1) $\rho_\eta^\kappa(\alpha)$ *is continuous and monotonically nondecreasing for $\alpha > 0$;*
2) $\lim_{\alpha \to \infty} \rho_\eta^\kappa(\alpha) = \|u_\delta\|^2 - \delta^2 - \left(\mu_\eta^\kappa(u_\delta, A_h) \right)^2;$
3) $\lim_{\alpha \downarrow 0} \rho_\eta^\kappa(\alpha) \leq -\delta^2;$
4) *if condition (1.8) is satisfied, then there is an $\alpha^* > 0$ such that*

$$\rho_\eta^\kappa(\alpha^*) = 0. \tag{1.19}$$

The latter equation is equivalent to (1.9), and the element $z_\eta^{\alpha^*}$ is not equal to zero and is uniquely defined.

PROOF. The assertions in 1)–3) above follow from the properties of the functions $\beta_\eta(\alpha)$, $\gamma_\eta(\alpha)$. It is clear that in 4) we only have to prove that $z_\eta^{\alpha^*}$ is unique.

Generally speaking, since $\rho_\eta^\kappa(\alpha)$ cannot be strictly monotone, α^* can be defined in a nonunique manner. Moreover, the set of roots of the equation $\rho_\eta^\kappa(\alpha^*) = 0$ fills, in general, a certain interval. Suppose α^* belongs to this interval. Then for all $z_\eta^{\alpha^*}$ condition (1.19) holds. This condition and the monotonicity of $\beta_\eta(\alpha)$, $\gamma_\eta(\alpha)$ imply that these functions are constant on interval mentioned above. Consequently, the element $z_\eta^{\alpha^*}$ associated with an α^* from the interval and determined in a unique manner, is an extremal of the functional (1.2) for any α^* satisfying (1.19). $\square$

The properties of the function $\rho_\eta(\alpha) = \beta_\eta(\alpha) - \left(\delta + h\sqrt{\gamma_\eta(\alpha)}\right)^2$ can easily be obtained in a similar manner; however, as already noted in §3, it is possible that the equation $\rho_\eta(\alpha) = 0$ has no root $\alpha > 0$.

Before turning to questions related with finding the root of the generalized discrepancy, we will consider what additional properties the auxiliary functions and the generalized discrepancy have in the case of a problem without constraints ($D = Z$) (or if z_η^α is an interior point of D).

In this case the extremal of the functional can be found as the solution to Euler's equation (1.5), to wit

$$z_\eta^a = R_\eta^\alpha A_h^* u_\delta = (A_h^* A_h + \alpha E)^{-1} A_h^* u_\delta.$$

LEMMA 1.4. The operator R_η^α has the following properties:

1) if $D(A_h) = Z$, then $D(R_\eta^\alpha) = Z$;
2) $\|R_\eta^\alpha\| \leq 1/\alpha$;
3) R_η^α is positive definite;
4) R_η^α depends continuously on α in the uniform operator topology.

PROOF. The operator $A_h^* A_h$ is selfadjoint and nonnegative definite, $\alpha > 0$, and so $\mathrm{Ker}(A_h^* A_h + \alpha E) = 0$. Whence R_η^α exists. We find its domain of definition. Let $R(A_h^* A_h + \alpha E)$ be the range of $A_h^* A_h + \alpha E$. This is clearly a linear manifold in Z. We show that it is closed. Suppose the sequence $z_n = (A_h^* A_h + \alpha E)\psi_n$ converges to $z_0 \in Z$. We show that there is a $\psi_0 \in Z$ such that $z_0 = (A_h^* A_h + \alpha E)\psi_0$.

Since

$$\psi_n = (A_h^* A_h + \alpha E)^{-1} z_n,$$
$$\|(A_h^* A_h + \alpha E)\psi_n\|^2 = \|A_h^* A_h \psi_n\|^2 + 2\alpha(A_h^* A_h \psi_n, \psi_n) + \alpha^2 \|\psi\|^2,$$
$$(A_h^* A_h \psi_n, \psi_n) \geq 0,$$

we see that

$$\|\psi_n\| \leq \frac{1}{\alpha}\|(A_h^* A_h + \alpha E)\psi_n\| \leq \frac{1}{\alpha}\|z_n\|. \tag{1.20}$$

Thus, the sequence $\{\psi_n\}$ is bounded. We extract a weakly convergent subsequence from it. Without loss of generality we may assume that $\psi_n \longrightarrow^{\mathrm{wk}} \psi_0$. Since the closed convex set $D(A_h^* A_h)$ in the Hilbert space Z is weakly closed, ψ_0 belongs to $D(A_h^* A_h)$. Since $A_h^* A_h + \alpha E$ is a bounded linear operator, it is continuous from the weak topology

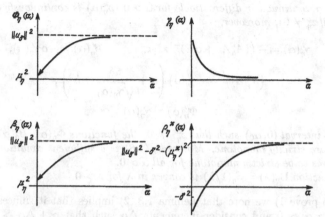

FIGURE 1.1. The functions $\Phi_\eta(\alpha)$, $\beta_\eta(\alpha)$, $\gamma_\eta(\alpha)$, and $\rho_\eta^\kappa(\alpha)$ when $D = Z$ and condition (1.8) holds.

into the weak topology [95], which implies that $(A_h^* A_h + \alpha E)\psi_0 = z_0$. Thus, we have proved that $R(A_h^* A_h + \alpha E)$ is a subspace of Z. We will prove that it coincides with Z. Write Z as a direct sum

$$Z = R(A_h^* A_h + \alpha E) \oplus \left(R(A_h^* A_h + \alpha E) \right)^\perp.$$

Assume that $z \in \left(R(A_h^* A_h + \alpha E) \right)^\perp$, i.e. $\left(z, (A_h^* A_h + \alpha E)y \right) = 0$ for all $y \in Z$. Taking $y = z$ we obtain $(z, A_h^* A_h z + \alpha z) = \|A_h z\|^2 + \alpha \|z\|^2 = 0$, i.e. $z = 0$. This means that $Z = R(A_h^* A_h + \alpha E)$ for all $\alpha > 0$.

The norm estimate for R_η^α given in 2) above follows directly from (1.20). The assertion in 4) is a consequence of the estimate in 2) and the relations

$$\|R_\eta^{\alpha + \Delta\alpha} - R_\eta^\alpha\| = \sup_{\|z\|=1} \|R_\eta^{\alpha+\Delta\alpha} z - R_\eta^\alpha z\| =$$

$$= \sup_{\|(A_h^* A_h + (\alpha+\Delta\alpha)E)\psi\|=1} \left\| \psi - \psi - \Delta\alpha (A_h^* A_h + \alpha E)^{-1}\psi \right\| \le$$

$$\le \frac{|\Delta\alpha|}{\alpha} \sup_{\|z\|=1} \left\| (A_h^* A_h + (\alpha + \Delta\alpha)E)^{-1} z \right\| \le \frac{|\Delta\alpha|}{\alpha(\alpha + \Delta\alpha)}.$$

This proves the lemma. □

Lemma 1.4 guarantees the solvability and the uniqueness of the solution of the Euler equation (1.5) for any $u_\delta \in U$. We will show what additional properties the functions $\rho_\eta^\kappa(\alpha)$, $\Phi_\eta(\alpha)$, $\beta_\eta(\alpha)$, $\gamma_\eta(\alpha)$ have in this case.

LEMMA 1.5. *Suppose* $D = Z$. *Then the functions* $\Phi_\eta(\alpha)$, $\beta_\eta(\alpha)$, $\gamma_\eta(\alpha)$, $\rho_\eta^\kappa(\alpha)$ *have the following properties, in addition to the properties listed in Lemma 1.3 and its corollaries (see Figure 1.1):*

1) *They are continuously differentiable for $\alpha > 0$ ($\rho_\eta^\kappa(\alpha)$ is continuously differentiable if $z_\eta^\alpha \neq 0$); moreover,*

$$\gamma_\eta'(\alpha) = - \left((A_h^* A_h + \alpha E)^{-1} z_\eta^\alpha, z_\eta^\alpha \right), \qquad \beta_\eta'(\alpha) = -\alpha \gamma_\eta'(\alpha),$$

$$\left(\rho_\eta^\kappa(\alpha) \right)' = -\gamma_\eta'(\alpha) \left(\alpha + \frac{h\delta}{\sqrt{\gamma_\eta(\alpha)}} + h^2 \right), \qquad (1.21)$$

$$\Phi_\eta''(\alpha) = \gamma_\eta'(\alpha).$$

2) *On an interval $(0, \alpha_0)$ such that $z_\eta^{\alpha_0} \neq 0$, the functions $\Phi_\eta(\alpha)$, $\beta_\eta(\alpha)$, $\gamma_\eta(\alpha)$, $\rho_\eta^\kappa(\alpha)$ are strictly monotone. Moreover, (1.8) is a sufficient condition for the functions to be strictly monotone for all $\alpha > 0$.*

3) *The function $B_\eta(\lambda) = \beta_\eta(1/\lambda)$ is convex in λ for $\lambda > 0$.*

PROOF. To prove 1) we note that Lemma 1.3, 2) implies that it suffices to find $\gamma_\eta'(\alpha)$. Fix an $\alpha > 0$ and consider increments $\Delta\alpha$ such that $\alpha + \Delta\alpha > 0$. Put $\Delta z_\eta^\alpha = z_\eta^{\alpha + \Delta\alpha} - z_\eta^\alpha$. By Lemma 1.4, 4), $\|\Delta z_\eta^\alpha\| \to 0$ as $\Delta\alpha \to 0$. The Euler equations corresponding to α and $\alpha + \Delta\alpha$ are:

$$A_h^* A_h z_\eta^\alpha + \alpha z_\eta^\alpha = A_h^* u_\delta,$$
$$A_h^* A_h (z_\eta^\alpha + \Delta z_\eta^\alpha) + (\alpha + \Delta\alpha)(z_\eta^\alpha + \Delta z_\eta^\alpha) = A_h^* u_\delta.$$

Subtracting these we obtain

$$\Delta z_\eta^\alpha = -\Delta\alpha \left(A_h^* A_h + (\alpha + \Delta\alpha) E \right)^{-1} z_\eta^\alpha.$$

Consider now the following difference relation:

$$\frac{\gamma_\eta(\alpha + \Delta\alpha) - \gamma_\eta(\alpha)}{\Delta\alpha} = \frac{\left(z_\eta^\alpha + \Delta z_\eta^\alpha, z_\eta^\alpha + \Delta z_\eta^\alpha \right) - \left(z_\eta^\alpha, z_\eta^\alpha \right)}{\Delta\alpha} =$$

$$= \frac{2 \left(\Delta z_\eta^\alpha, z_\eta^\alpha \right) + \left(\Delta z_\eta^\alpha, \Delta z_\eta^\alpha \right)}{\Delta\alpha} =$$

$$= -\frac{2\Delta\alpha}{\Delta\alpha} \left(\left(A_h^* A_h + (\alpha + \Delta\alpha) E \right)^{-1} z_\eta^\alpha, z_\eta^\alpha \right) +$$

$$+ \frac{(\Delta\alpha)^2}{\Delta\alpha} \left\| \left(A_h^* A_h + (\alpha + \Delta\alpha) E \right)^{-1} z_\eta^\alpha \right\|^2.$$

Consequently,

$$\frac{\gamma_\eta(\alpha + \Delta\alpha) - \gamma_\eta(\alpha)}{\Delta\alpha} \longrightarrow -2 \left(\left(A_h^* A_h + (\alpha+) E \right)^{-1} z_\eta^\alpha, z_\eta^\alpha \right)$$

as $\Delta\alpha \to 0$. Here we have used the continuity of R_η^α with respect to α and the boundedness of R_η^α as a linear operator from Z into Z. Formula (1.21) has been proved.

Since R_η^α is a positive definite operator (Lemma 1.4), $\gamma_\eta'(\alpha) < 0$ for all $\alpha \in (0, \alpha_0]$, where $z_\eta^{\alpha_0} \neq 0$, i.e. $\gamma_\eta(\alpha)$ (and also $\Phi_\eta(\alpha)$, $\beta_\eta(\alpha)$, $\rho_\alpha^\kappa(\alpha)$) is strictly monotone on $(0, \alpha_0]$.

To complete the proof of 2) above, we note that since U is a Hilbert space, it can be written as a direct sum of orthogonal subspaces [90]:

$$U = \overline{A_h Z} \oplus \operatorname{Ker} A_h^*.$$

Therefore u_δ can be uniquely written as $u_\delta = v_\delta + w_\delta$, $v_\delta \in \operatorname{Ker} A_h^*$, $w_\delta \in \overline{A_h Z}$. Moreover,

$$\left(\mu_\eta(u_\delta, A_h)\right)^2 = \inf_{z \in Z} \|A_h z - u_\delta\|^2 = \|w_\delta - u_\delta\|^2 = \|v_\delta\|^2.$$

If $z_\eta^\alpha = 0$, then Euler's equation implies that $A_h^* u_\delta = 0$, i.e. $u_\delta \in \operatorname{Ker} A_h^*$, $u_\delta = v_\delta$. Consequently, if $\|u_\delta\|^2 > \delta^2 + \left(\mu_\eta(u_\delta, A_h)\right)^2 = \delta^2 + \left(\|v_\delta\| + \kappa\right)^2$, then z_η^α cannot be equal to zero for any $\alpha > 0$.

Item 3) can be proved by the change of variable $\alpha = 1/\lambda$ and a direct computation of the second derivative of the function $\beta_\eta(1/\lambda)$ in a manner similar to the computation of the first derivative of $\gamma_\eta(\alpha)$. $\square$

COROLLARY. *If $D = Z$, then α^*, defined in accordance with the generalized discrepancy principle $\rho_\eta^\kappa(\alpha^*) = 0$, is unique.*

REMARK. For the finite-dimensional case the derivatives of $\Phi_\eta(\alpha)$, $\gamma_\eta(\alpha)$ and $\beta_\eta(\alpha)$ have been computed in [135], [136] using the expansion of the solution of the Euler equation (1.5) in eigenvectors of the matrix $A_h^* A_h$. In these papers it was also noted that $\beta_\eta(\lambda) = \beta_\eta(1/\lambda)$ is a convex function of λ, which made it possible to solve by Newton's method the equation $\beta_\eta(\lambda) = \delta^2$ when choosing the regularization parameter in accordance with the discrepancy principle ($h = 0$).

We will consider in some detail the problem of finding the root of the equation (1.19) in case $D = Z$. In this case $\rho_\eta^\kappa(\alpha)$ is a strictly monotone, increasing (if (1.8) holds), differentiable function for $\alpha > 0$, and if $h = 0$ the function $\sigma_\eta^\kappa(\lambda) = \rho_\eta^\kappa(1/\lambda)$ is convex. For a successful search of the root of (1.19) we would like to have an upper bound for the regularization parameter α^*, i.e. a parameter value $\overline{\alpha} \geq \alpha^*$ such that $\rho_\eta^\kappa(\overline{\alpha}) > 0$. Note that in [56] an upper bound for the regularization parameter chosen in accordance with the discrepancy principle ($h = 0$) has been obtained for equations of convolution type; this was later generalized in [38].

For simplicity we assume that $\overline{A_h Z} = U$, i.e. $\mu_\eta^\kappa(u_\delta, A_h)$ can be regarded as being zero (the case $\mu_\eta^\kappa(u_\delta, A_h) > 0$ can be treated completely similar, by replacing δ with $\delta + \mu_\eta^\kappa(u_\delta, A_h)$).

LEMMA 1.6 ([38]). *Suppose $D = Z$, $z_\eta^\alpha \neq 0$. Then the following inequalities hold:*

$$\alpha \leq \frac{\|A_h\| \|A_h z_\eta^\alpha - u_\delta\|}{\|z_\eta^\alpha\|}, \tag{1.22}$$

$$\alpha \leq \frac{\|A_h\|^2 \|A_h z_\eta^\alpha - u_\delta\|}{\|u_\delta\| - \|A_h z_\eta^\alpha - u_\delta\|}. \tag{1.23}$$

PROOF. Since z_η^α is an extremal of the functional $M^\alpha[z]$, this functional attains on the elements $(1 - \gamma)z_\eta^\alpha$, $\gamma \in (-\infty, +\infty)$, its minimum for $\gamma = 0$. Consequently, for any $\gamma \in (-\infty, +\infty)$:

$$0 \leq M^\alpha[(1 - \gamma)z_\eta^\alpha] - M^\alpha[z_\eta^\alpha].$$

We estimate $M^\alpha[(1 - \gamma)z_\eta^\alpha]$ from below for $\gamma \in [0, 1]$:

$$M^\alpha[(1 - \gamma)z_\eta^\alpha] = \|A_h z_\eta^\alpha - u_\delta - \gamma A_h z_\eta^\alpha\|^2 + \alpha(1 - \gamma)^2\|z_\eta^\alpha\|^2 \leq$$
$$\leq \|A_h z_\eta^\alpha - u_\delta\|^2 + \alpha\|z_\eta^\alpha\|^2 + 2\gamma\left(\|A_h z_\eta^\alpha - u_\delta\|\,\|A_h z_\eta^\alpha\| - \alpha\|z_\eta^\alpha\|^2\right) +$$
$$+ \gamma^2\left(\|A_h z_\eta^\alpha\|^2 + \alpha\|z_\eta^\alpha\|^2\right).$$

Hence,

$$0 \leq M^\alpha[(1 - \gamma)z_\eta^\alpha] - M^\alpha[z_\eta^\alpha] \leq$$
$$\leq 2\gamma\left(\|A_h z_\eta^\alpha - u_\delta\|\,\|A_h z_\eta^\alpha\| - \alpha\|z_\eta^\alpha\|^2\right) + \gamma^2\left(\|A_h z_\eta^\alpha\|^2 + \alpha\|z_\eta^\alpha\|^2\right).$$

Dividing this inequality by $\gamma > 0$ and taking the limit as $\gamma \to 0$ we obtain

$$\alpha \leq \frac{\|A_h z_\eta^\alpha\|\,\|A_h z_\eta^\alpha - u_\delta\|}{\|z_\eta^\alpha\|^2},$$

which readily implies (1.22).

It remains to note that

$$\|z_\eta^\alpha\| \geq \frac{\|A_h z_\eta^\alpha\|}{\|A_h\|} \geq \frac{\|u_\delta\| - \|A_h z_\eta^\alpha - u_\delta\|}{\|A_h\|}$$

and that $\|u_\delta\| \geq \|A_h z_\eta^\alpha - u_\delta\|$, since otherwise we would have $M^\alpha[0] \leq M^\alpha[z_\eta^\alpha]$, contradicting the condition $z_\eta^\alpha \neq 0$. Substituting the lower bound for z_η^α into (1.22) we obtain (1.23). □

COROLLARY. *If*

$$\tilde{\alpha} = \frac{\|A_h\|^2 C\delta}{\|u_\delta\| - C\delta}, \qquad C = \text{const} \geq 1, \tag{1.24}$$

then $\beta_\eta(\tilde{\alpha}) \geq C^2\delta^2$.

In fact, if $\beta_\eta(\alpha) = C^2\delta^2$, then, substituting $\|A_h z_\eta^\alpha - u_\delta\| = C\delta$ into (1.23), we obtain an upper bound $\tilde{\alpha}$ for the regularization parameter chosen in accordance with the discrepancy principle ($h = 0$). If $C = 1$, then the equation $\beta_\eta(\alpha) = \delta^2$ can have a solution also for $\alpha = 0$ [36], which shows that there is no lower bound for α if no additional assumptions are made.

LEMMA 1.7 ([202]). *Suppose that the exact solution of* (1.1) *is* $\overline{z} \neq 0$, *that* $\|u_\delta - A_h\overline{z}\| \leq \delta$ *and* $\|u_\delta\|/\delta > C > 1$, $C = \text{const}$, *for all* $\delta \in (0, \delta_0]$. *Then* $\alpha^* \leq \overline{\alpha}$, *where*

$$\overline{\alpha} = \|A_h\|\left(h + \frac{\sqrt{h^2 + (\tilde{\alpha} + h^2)(C^2 - 1)} + h}{C^2 - 1}\right);$$

here, $\tilde{\alpha}$ *is defined by* (1.24).

PROOF. (1.22) implies

$$\alpha^* \leq \frac{\|A_h\| \, \|A_h z_\eta^{\alpha^*} - u_\delta\|}{\|z_\eta^{\alpha^*}\|} = \|A_h\| \frac{\delta + h\|z_\eta^{\alpha^*}\|}{\|z_\eta^{\alpha^*}\|} = \|A_h\| \left(h + \frac{\delta}{\|z_\eta^{\alpha^*}\|} \right).$$

To estimate $\|z_\eta^{\alpha^*}\|$ from below we use the fact that $z_\eta^{\tilde\alpha}$ is an extremal of $M^{\tilde\alpha}[z]$:

$$0 < C^2\delta^2 \leq \beta_\eta(\tilde\alpha) \leq \|A_h z_\eta^{\tilde\alpha} - u_\delta\|^2 + \tilde\alpha\|z_\eta^{\tilde\alpha}\|^2 \leq$$
$$\leq \|A_h z_\eta^{\alpha^*} - u_\delta\|^2 + \tilde\alpha\|z_\eta^{\alpha^*}\|^2 = (\delta + h\|z_\eta^{\alpha^*}\|)^2 + \tilde\alpha\|z_\eta^{\alpha^*}\|^2 =$$
$$= (h^2 + \tilde\alpha)\|z_\eta^{\alpha^*}\|^2 + 2h\delta\|z_\eta^{\alpha^*}\| + \delta^2.$$

This implies that

$$\|z_\eta^{\alpha^*}\| \geq \delta \frac{C^2 - 1}{\sqrt{h^2 + (C^2 - 1)(\tilde\alpha + h^2)} + h}.$$

Substituting the bound for $\|z_\eta^{\alpha^*}\|$ into the bound for α^*, we arrive at the assertion of the lemma. $\square$

We will consider in more detail how to use Newton's method for solving equation (1.19). In general, Newton's method can be used only if the initial approximation is sufficiently close to the root α^* of (1.19) (since $(\rho_\eta^\kappa(\alpha))' > 0$ and $(\rho_\eta^\kappa(\alpha))''$ exists and is continuous for $\alpha > 0$, [91]). The function $\sigma_\eta^\kappa(\lambda) = \rho_\eta^\kappa(1/\lambda)$ is convex with respect to λ if $h = 0$, hence Newton's iteration process can be constructed as follows:

$$\lambda_{n+1} = \lambda_n - \frac{\sigma_\eta^\kappa(\lambda_n)}{\left(\sigma_\eta^\kappa(\lambda_n)\right)'}, \qquad n = 0, 1, \ldots,$$

or

$$\alpha_{n+1} = \frac{\alpha_n^2 \left(\rho_\eta^\kappa(\alpha_n)\right)'}{\alpha_n + \rho_\eta^\kappa(\alpha_n)}, \qquad n = 0, 1, \ldots. \tag{1.25}$$

As initial approximation α_0 we can take the $\overline{\alpha}$ given by Lemma 1.7. For $h = 0$ the sequence α_n given by (1.25) converges to α^* (in this case we can naturally take α_0 equal to the $\tilde\alpha$ from (1.24) for $C = 1$). For $h \neq 0$ the convergence of this iteration process is not guaranteed. To find a solution of (1.19) in this case, we recommend the use of the method of dividing the interval into halves or the chord method [9], or a combination of these methods with Newton's method. Methods for finding a root (provided it exists) of the generalized discrepancy $\rho_\eta(\alpha) = \|A_h z_\eta^\alpha - u_\delta\|^2 - \left(\delta + h\|z_\eta^\alpha\|\right)^2$ are similar.

Note that for searching an extremal of the smoothing functional (or of its finite-difference approximation) for a fixed $\alpha > 0$ we may use, next to the Euler equation, some method for minimizing a differentiable convex functional in a Hilbert space (the method of steepest descent, of conjugate gradients, Newton's method, etc.). A detailed description of these methods can be found in, e.g., [32], [33], [93], [144], [146], [152], [191], [192].

We will now consider the case $D \neq Z$. In this case $\gamma_\eta(\alpha)$, $\beta_\eta(\alpha)$, $\rho_\eta^\kappa(\alpha)$ are not, in general, differentiable for all $\alpha > 0$. The regularization parameter can be chosen in accordance with the generalized discrepancy principle, i.e. by solving (1.19), but to find the root of (1.19) we have to use numerical methods that do not require us to compute derivatives of $\rho_\eta^\kappa(\alpha)$. Such methods include, e.g., the chord method and the method of dividing the interval into halves. Also, to find $z_\eta^\alpha \in D$ for a fixed $\alpha > 0$ we have to use gradient methods for minimizing the smoothing functional with constraints [43], [145].

Below we will consider certain algorithms for the numerical realization of the generalized discrepancy principle for problems both without or with simple constraints.

In this book we will completely avoid the problem of solving unstable extremal problems, which is considered in, e.g., [14], [28], [33], [111], [117], [137], [171]–[173], [179]. We will also pass by problems of using regularized Newton, quasiNewton and other iteration methods [1], [15], [16], [30], [31], [44], [45], [80], [81], [154], [187], [188].

5. Finite-dimensional approximation of ill-posed problems

To solve ill-posed problems it is usually necessary to approximate the initial, often infinite-dimensional, problem by a finite-dimensional problem, for which numerical algorithms and computer programs have been devised.

Here we will consider the problem of passing to a finite-dimensional approximation by using the example of a Fredholm integral equation of the first kind. We will not dwell on the additional conditions to be imposed so as to guarantee convergence of the finite-dimensional extremals of the approximate smoothing functional to extremals of the initial functional $M^\alpha[z]$ when the dimension of the finite-dimensional problem increases unboundedly [90]. Transition to a finite-dimensional approximation can be regarded as the introduction of an additional error in the operator and we may use a modification of the generalized discrepancy principle [61]. In the sequel we will assume that the dimension of the finite-dimensional problem is chosen sufficiently large, so that the error of approximating the operator A in (1.1) is substantially smaller than the errors h and δ.

Consider the *Fredholm integral equation of the first kind*

$$Az = \int_a^b K(x,s)z(s)\,ds = u(x), \qquad c \le x \le d. \tag{1.26}$$

We will assume that $K(x,s)$ is a real-valued function defined and continuous on the rectangle $\Pi = \{a \le s \le b, c \le x \le d\}$. For simplicity reasons, we will also assume that the kernel K is nonsingular. Instead of $\overline{u} = A\overline{z}$ we know an approximate value u_δ such that $\|u_\delta - \overline{u}\|_{L_2} \le \delta$, i.e. $U = L_2[c,d]$. Suppose that we may conclude from a priori considerations that the exact solution $\overline{z}(s)$ corresponding to $\overline{u}(x)$ is a smooth function on $[a,b]$. For example, we will assume that $\overline{z}(s)$ is continuous on $[a,b]$ and has almost everywhere a derivative which is square-integrable on $[a,b]$. In this case we may naturally take $Z = \mathcal{W}_2^1[a,b]$.

Suppose that instead of $K(x,s)$ we know a function $K_h(x,s)$ such that $\|K - K_h\|_{L_2(\Pi)} \le h$. Then $\|A - A_h\|_{\mathcal{W}_2^1 \to L_2} \le h$, where A_h is the integral operator with kernel $K_h(x,s)$.

Using the standard scheme for constructing a regularizing algorithm, we obtain an approximate solution $z_\eta^{\alpha(\eta)} \in Z = \mathcal{W}_2^1[a,b]$ which converges, as $\eta \to 0$, to $\overline{z}$ in the norm of the space $\mathcal{W}_2^1[a,b]$. The Sobolev imbedding theorem [153] implies that $z_\eta^{\alpha(\eta)}$ converges uniformly on $[a,b]$ to $\bar{z}$ as $\eta \to 0$, i.e.

$$\max_{a \le s \le b} |z_\eta^{\alpha(\eta)}(s) - \bar{z}(s)| \longrightarrow^{\eta \to 0} 0.$$

In this setting the functional $M^\alpha[z]$ for the problem (1.26) takes the form

$$M^\alpha[z] = \|A_h z - u_\delta\|_{L_2}^2 + \alpha\|z\|_{\mathcal{W}_2^1}^2 =$$
$$= \int_c^d \left[\int_a^b K_h(x,s) z(s)\, ds - u_\delta(x) \right]^2 dx + \alpha \int_a^b \left\{ z^2(s) + [z'(s)]^2 \right\} ds. \tag{1.27}$$

In the construction of a finite-difference approximation, for simplicity we will use the expression (1.27) for the smoothing functional.

To this end we first of all choose grids $\{s_j\}_{j=1}^n$ and $\{x_i\}_{i=1}^m$ on the intervals $[a,b]$ and $[c,d]$, respectively. Then, using quadrature formulas, e.g. those of the trapezium method, we can construct finite-difference analogs of the operator A in the integral equation (1.26). The finite-difference operator (which we will denote also by $A: \mathbf{R}^n \to \mathbf{R}^m$ if this doesn't lead to confusion) is a linear operator with matrix $A = \{a_{ij}\}$. The simplest version of approximation that we will use in the sequel is that given by the formulas

$$\left. \begin{array}{ll} a_{ij} = K_h(x_i, s_j), & j = 2, \ldots, n-1, \\[2mm] a_{ij} = \dfrac{K_h(x_i, s_j)}{2}, & j = 1, n, \end{array} \right\} \quad i = 1, \ldots, m.$$

To complete the transition to the finite-dimensional problem it now suffices to approximate the integrals occurring in the mean-square norms of the spaces L_2 and $\mathcal{W}_2^1$. For simplicity we will assume that the grids are uniform with steps h_s and h_x. We put $z(s_j) = z_j$, $u_\delta(x_i) = u_i$. Using the rectangle formula to approximate the integrals, we obtain

$$\int_c^d \left[\int_a^b K_h(x,s) z(s)\, ds - u_\delta(x) \right]^2 dx \approx \sum_{i=1}^m \left[\sum_{j=1}^n a_{ij} z_j h_s - u_i \right]^2 h_x,$$
$$\int_a^b [z(s)]^2\, ds \approx \sum_{j=1}^n z_j^2 h_s, \qquad \int_a^b [z'(s)]^2\, ds \approx \sum_{j=2}^n \frac{(z_j - z_{j-1})^2}{h_s}.$$

It is now easy to describe the conditions for a minimum, with respect to the variables z_j, $j = 1, \ldots, n$, of the functional

$$\hat{M}^\alpha[z] = \sum_{i=1}^m \left[\sum_{j=1}^n a_{ij} z_j h_s - u_i \right]^2 h_x + \alpha \sum_{j=1}^n z_j^2 h_s + \alpha \sum_{j=2}^n \frac{(z_j - z_{j-1})^2}{h_s}$$

approximating (1.27):

$$h_x h_s \sum_{k=1}^{n} \left[\sum_{i=1}^{m} a_{ik} a_{i1} \right] z_k + \alpha h_s z_1 - \alpha \frac{z_2 - z_1}{h_s} = \sum_{i=1}^{m} a_{i1} u_i h_x h_s;$$

$$h_x h_s \sum_{k=1}^{n} \left[\sum_{i=1}^{m} a_{ik} a_{ij} \right] z_k + \alpha h_s z_j - \alpha \frac{z_{j-1} - 2z_j + z_{j-1}}{h_s} = \sum_{i=1}^{m} a_{ij} u_i h_x h_s;$$

$$j = 2, \ldots, n-1,$$

$$h_x h_s \sum_{k=1}^{n} \left[\sum_{i=1}^{m} a_{ik} a_{in} \right] z_k + \alpha h_s z_n - \alpha \frac{z_{n-1} - z_n}{h_s} = \sum_{i=1}^{m} a_{in} u_i h_x h_s.$$

Setting $h_x \sum_{i=1}^{n} a_{ik} a_{ij} = b_{jk}$, the b_{jk} become the entries of a matrix B, and setting $\sum_{i=1}^{m} a_{ij} u_i h_x = f_j$, the f_j become the components of a vector f. Thus, we arrive at the problem of solving the system of equations

$$B^\alpha z = Bz + \alpha Cz = f, \tag{1.28}$$

where

$$C = \begin{pmatrix} 1 + \frac{1}{h_s^2} & -\frac{1}{h_s^2} & 0 & \cdots & 0 & 0 \\ -\frac{1}{h_s^2} & 1 + \frac{2}{h_s^2} & -\frac{1}{h_s^2} & \cdots & 0 & 0 \\ \cdots & \cdots & \cdots & \cdots & \cdots & \cdots \\ 0 & 0 & 0 & \cdots & -\frac{1}{h_s^2} & 1 + \frac{1}{h_s^2} \end{pmatrix}.$$

Note that if we regard the operator A in the integral equation (1.26) as an operator from $L_2[a, b]$ into $L_2[c, d]$ (when information regarding the smoothness of the exact solution $\overline{z}(s)$ is absent), then the smoothing functional has the form

$$M^\alpha[z] = \|A_h z - u_\delta\|_{L_2}^2 + \alpha \|z\|_{L_2}^2,$$

and the equation for its extremum after transition to the finite-difference approximation can be written in the form (1.28) with matrix

$$C = E = \begin{pmatrix} 1 & 0 & 0 & \cdots & 0 & 0 \\ 0 & 1 & 0 & \cdots & 0 & 0 \\ \cdots & \cdots & \cdots & \cdots & \cdots & \cdots \\ 0 & 0 & \cdots & 0 & 0 & 1 \end{pmatrix}.$$

We can arrive at the system (1.28) by starting with the Euler equation $A_h^* A_h z - A_h^* u_\delta + \alpha z = 0$ in $W_2^1[a, b]$. Here, A_h is an operator from $W_2^1[a, b]$ into $L_2[c, d]$ and A_h^* is its adjoint, $A_h^*: L_2[c, d] \rightarrow W_2^1[a, b]$. Using the properties of A_h^* [165], [181] and passing from the Euler equation to the finite-difference approximation, it is not difficult to obtain the system of equations (1.28).

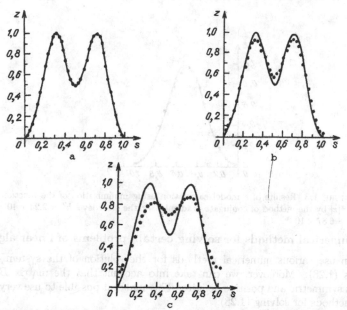

FIGURE 1.2. Results of a model calculation using a solution of the Euler equation and with regularization parameter chosen in accordance with the generalized discrepancy principle, for the following error levels: a) $h^2 = 10^{-10}$, $\delta^2 = 10^{-8}$; b) $h^2 = 2.24 \times 10^{-7}$, $\delta^2 = 6.41 \times 10^{-6}$; c) $h^2 = 2.24 \times 10^{-7}$, $\delta^2 = 3.14 \times 10^{-4}$.

As an illustration we consider some results of solving a model problem for equation (1.26). Let

$$a = 0, \quad b = 1, \quad c = -2, \quad d = 2, \quad n = m = 41,$$

$$K(x,s) = \frac{1}{1 + 100(x - s)^2},$$

$$\overline{z}(s) = \frac{\exp\left\{-\frac{(s-0.3)^2}{0.03}\right\} + \exp\left\{-\frac{(s-0.7)^2}{0.03}\right\}}{10.9550408} - 0.052130913.$$

In Figure 1.2 we have given the results, for various error levels, of computer calculations using a numerical solution of the Euler equation and with regularization parameter chosen in accordance with the generalized discrepancy principle ($\overline{z}(s)$ is represented by the continuous line; the approximate solution by points).

To find the minimum of $\hat{M}^\alpha[z]$ we can use a numerical method for minimizing functionals, e.g. the method of conjugate gradients. Here we may consider problems with constraints ($D \neq Z$). See Chapter 3 for more about the use of gradient methods; here we only give the results of the model calculation. Let $\overline{z}(s) = \exp\{-(s - 0.5)^2\}/0.06\}$, and let $K(x,s)$ and other parameters be as above. See Figure 1.3 for the results.

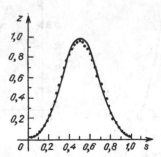

FIGURE 1.3. Results of a model calculation using minimization of the functional $\hat{M}^\alpha[z]$ by the method of conjugate gradients, for the error level $h^2 = 2.24 \times 10^{-7}$, $\delta^2 = 6.87 \times 10^{-6}$.

6. Numerical methods for solving certain problems of linear algebra

We can use various numerical methods for the solution of the system of linear equations (1.28). Moreover, we can take into account that the matrix B^α of the system is symmetric and positive definite. This makes it possible to use very efficient special methods for solving (1.28).

The square-root method has been proposed as one such method [189]. Since the entries b_{ij} of B^α are real numbers, and B^α is symmetric and positive definite, B^α can be written as the product of matrices $(T^\alpha)^*T^\alpha$ with T^α an upper triangular matrix:

$$T^\alpha = \begin{pmatrix} t_{11}^\alpha & t_{12}^\alpha & \cdots & t_{1n}^\alpha \\ 0 & t_{22}^\alpha & \cdots & t_{2n}^\alpha \\ \cdots & \cdots & \cdots & \cdots \\ 0 & 0 & \cdots & t_{nn}^\alpha \end{pmatrix}.$$

The entries of T^α can be successively found by the formulas

$$t_{11}^\alpha = \sqrt{b_{11}^\alpha}, \qquad t_{1j}^\alpha = \frac{b_{1j}^\alpha}{t_{11}^\alpha}, \quad j = 2, \ldots, n,$$

$$t_{ii}^\alpha = \left\{ b_{ii}^\alpha - \sum_{k=1}^{i-1} (t_{ki}^\alpha)^2 \right\}^{1/2}, \qquad i = 2, \ldots, n, \tag{1.29}$$

$$t_{ij}^\alpha = \frac{b_{ij}^\alpha - \sum_{k=1}^{i-1} t_{ki}^\alpha t_{kj}^\alpha}{t_{ii}^\alpha}, \quad i < j, \qquad t_{ij}^\alpha = 0, \quad i > j.$$

The system (1.28) takes the form

$$(T^\alpha)^* T^\alpha z^\alpha = f.$$

Introducing the notation $y^\alpha = T^\alpha z^\alpha$, we can replace (1.28) by the two equivalent equations

$$(T^\alpha)^* y^\alpha = f, \qquad t^\alpha z^\alpha = y^\alpha.$$

Each of these equations can be elementary solved, since each involves a triangular matrix. Economic standard programs for solving a system of linear equations by the square root method have been given in [189].

To find the roots of the discrepancy or generalized discrepancy we will repeatedly solve the system of equations (1.28), for various $\alpha > 0$. Here, the matrix B^α of (1.28) depends on α in a special manner, while the righthand side does not change at all. This makes it possible to construct special economic methods for repeatedly solving (1.28) (see [41]).

Suppose we have to solve the system of equations

$$(A_h^* A_h + \alpha C) z^\alpha = A_h^* u$$

for various $\alpha > 0$. Here, A_h is a real matrix of order $m \times n$, $z^\alpha \in \mathbf{R}^n$, $u \in \mathbf{R}^m$, $m \geq n$, C is a positive definite symmetric matrix, A_h^* is the transposed matrix of A_h.

Using the square root method, the tridiagonal matrix C can be written, using (1.29) as $C = S^* S$ (note that S is bidiagonal). Changing to $y^\alpha = S z^\alpha$ ($z^\alpha = S^{-1} y^\alpha$), we obtain

$$(A_h^* A_h + \alpha C) S^{-1} y^\alpha = A_h^* u.$$

Multiplying the lefthand side by $(S^{-1})^*$, we obtain

$$(D^* D + \alpha E) y^\alpha = D^* u, \qquad D = A_h S^{-1}.$$

Write D as $D = QPR$ with Q an orthogonal matrix of order $m \times m$, R an orthogonal matrix of order $n \times n$, and P a right bidiagonal matrix of order $m \times n$ (in P only the entries p_{ii}, $p_{i,i+1}$ are nonzero). To construct such a decomposition it suffices to find Q^{-1}, R^{-1} such that $P = Q^{-1} D R^{-1}$ is bidiagonal.

For example, the matrices Q^{-1}, R^{-1} can be looked for in the form $Q^{-1} = Q_n \ldots Q_1$, $R^{-1} = R_1 \ldots R_n$, where Q_i, R_i are the matrices of reflection operators ($i = 1, \ldots, n$; see [40]), satisfying $Q_i = Q_i^* = Q_i^{-1}$, $R_i = R_i^* = R_i^{-1}$ ($i = 1, \ldots, n$). Then, $Q = Q_1 \ldots Q_n$, $R = R_n \ldots R_1$.

We will construct Q_i, R_i as follows. Let a_1 be the first column of D. We will look for Q_1 satisfying the requirement that all first column entries, from the second onwards, of the matrix $Q_1 D$ vanish, i.e. if q_j are the rows of Q_1, then $(q_j, a_1) = 0$, $j = 2, \ldots, m$; $q_j \in \mathbf{R}^m$. The matrices satisfying this condition are the matrices of reflection operators [40] with generating column vector

$$g^{(1)} = \frac{a_1 - \|a_1\| l}{\|a_1 - \|a_1\| l\|},$$

where l is the column vector in $\mathbf{R}^m$ with coordinates $(1, 0, \ldots, 0)$. Thus,

$$Q_1 = E - 2 g^{(1)} \left(g^{(1)} \right)^*.$$

We will now choose R_1 such that for the matrix $(Q_1 D) R_1$, first, all entries in the first column from the second entry onwards are nonzero, and, secondly, all entries in the first row from the third entry onwards are zero. The first requirement is satisfied if R_1 is looked for in the form

$$R_1 = \begin{pmatrix} 1 & 0 & \cdots & 0 \\ 0 & & & \\ \vdots & & \tilde{R}_1 & \\ 0 & & & \end{pmatrix}.$$

Let $\tilde{b}_1$ be the first row of $Q_1 D$ without the first element, $\tilde{b}_1 \in \mathbf{R}^{n-1}$. Then the second requirement means that $(\tilde{b}_1, v_i) = 0$, where v_i are the columns of $\tilde{R}_1$, $v_i \in \mathbf{R}^{n-1}$ $(i = 3, \ldots, n)$. Hence, $\tilde{R}_1$ is a reflection matrix, with generating vector

$$\tilde{h} = \frac{\tilde{b}_1 - \|\tilde{b}_1\| l}{\left\| \tilde{b}_1 - \|\tilde{b}_1\| l \right\|} \in \mathbf{R}^{n-1}.$$

But now R_1 is the reflection matrix with generating vector $\begin{pmatrix} 0 \\ \tilde{h} \end{pmatrix} = h^{(1)} \in \mathbf{R}^n$.

Further, we can similarly look for Q_i, R_i in the form

$$Q_i = \begin{pmatrix} E^{(i-1)} & 0 \\ 0 & \tilde{Q}_i \end{pmatrix}, \qquad R_i = \begin{pmatrix} E^{(i)} & 0 \\ 0 & \tilde{R}_i \end{pmatrix},$$

where $\tilde{Q}_i$, $\tilde{R}_i$ are reflection matrices in spaces of lower dimension. Note that to find Q and R we need not multiply out the matrices Q_i and R_i. It suffices to know the generating vectors $g^{(i)}$ and $h^{(i)}$, and the actions of Q_i and R_i on a vector w can be computed by

$$Q_i w = w - 2g^{(i)}(g^{(i)}, w), \qquad R_i w = w - 2h^{(i)}(h^{(i)}, w).$$

So, assume we have found matrices P, Q, R such that $D = QPR$. We now make the change of variables $x^\alpha = R y^\alpha$ $(y^\alpha = R^{-1} x^\alpha)$ in $(D^* D + \alpha E) y^\alpha = D^* u$. We obtain $(R^* P^* Q^* Q P R + \alpha E) R^{-1} x^\alpha = D^* u$, or $(P^* P + \alpha E) x^\alpha = R D^* u = f$. The matrix $P^* P$ is tridiagonal, and the latter equation can be solved without difficulty by using nth order operations, e.g. by the sweep method [151]. The operator $S^{-1} R^{-1}$ realizes the inverse transition from x^α to z^α. However, often we need not carry out this transition to z^α for all α. For example, if $h = 0$ and we choose α in accordance with the discrepancy principle, then we only have to verify the condition $\|A_h z^\alpha - u\| = \delta$, which is equivalent to the condition $\|P x^\alpha - Q^* u\| = \delta$, since $\|P x^\alpha - Q^* u\| = \|A_h z^\alpha - u\|$.

In Chapter 4 we consider programs implementing the algorithm described above.

7. Equations of convolution type

Even when solving one-dimensional Fredholm integral equations of the first kind on large computers, for each variable the dimension of the grid may not exceed 80—100 points. For equations of convolution type, which we will consider below, in the one-dimensional case it turns out to be possible to construct numerical solution methods with grids of more than 1000 points, using only the operating memory of a computer of average capacity. Here we use the specific form of the equations of convolution type and apply the Fourier transform (for certain other types of equations of the first kind with kernels of a special form it may be more efficient to use other integral transforms, see [82], [143], [178]). The development of numerical methods especially tuned to equations of convolution type and of the first kind started in [53], [56], [156]–[158], [178].

In this section we will consider methods for solving one- and two-dimensional equations of convolution type. We will consider integral equations of the first kind

$$Az = \int_{-\infty}^{+\infty} K(x-s)z(s)\,ds = u(x), \qquad -\infty < x < +\infty, \qquad (1.30)$$

which are often met in practice (for examples in problems of physics see [71], [178]). Suppose the functions in this equation satisfy the requirements

$$K(y) \in L_1(-\infty, +\infty) \cap L_2(-\infty, +\infty),$$
$$u(x) \in L_2(-\infty, +\infty),$$
$$z(s) \in \mathcal{W}_2^1(-\infty, +\infty),$$

i.e. $A \colon W_2^1 \to L_2$. We will also assume that the kernel $K(y)$ is closed, i.e. A is a bijective operator. Equation (1.30) is regarded without constraints ($D = Z$). Suppose that a function $\overline{u}(x)$ gives rise to a unique solution $\overline{z}(s) \in \mathcal{W}_2^1$ of (1.30). Suppose also that we do not know $\overline{u}(x)$ and A themselves, but only a function $u_\delta(x)$ and an operator A_h of convolution type with kernel $K_h(y)$ such that

$$\|u_\delta - \overline{u}\|_{L_2} \le \delta, \qquad \|A - A_h\|_{W_2^1 \to L_2} \le h,$$

where $\delta > 0$ and $h > 0$ are known errors. Consider the smoothing functional

$$M^\alpha[z] = \|A_h z - u_\delta\|_{L_2}^2 + \alpha\|z\|_{\mathcal{W}_2^1}^2.$$

Since $\mathcal{W}_2^1(-\infty, +\infty)$ is a Hilbert space, for any $\alpha > 0$ and $u_\delta \in L_2$ there is (see §1) a unique element $z_\eta^\alpha(s)$ realizing the minimum of $M^\alpha[z]$. If we choose the regularization parameter in accordance with the generalized discrepancy principle (see §2), then $z_\eta^{\alpha(\eta)}(s)$ tends, in the norm of $\mathcal{W}_2^1$, to the exact solution of (1.30) as $\eta \to 0$. For any interval $[a, b]$ the space $\mathcal{W}_2^1[a, b]$ can be compactly imbedded into $C[a, b]$ (see [153]), therefore $z_\eta^{\alpha(\eta)}(s)$ converges uniformly to $\overline{z}(s)$ on every closed interval on the real axis.

Using the convolution theorem, the Plancherel equality [23] and varying $M^\alpha[z]$ over $\mathcal{W}_2^1$, we obtain [3], [178]:

$$z_\eta^\alpha(s) = \frac{1}{2\pi} \int_{-\infty}^{+\infty} \frac{\tilde{K}_h^*(\omega)\tilde{u}_\delta(\omega)e^{-i\omega s}}{L(\omega) + \alpha(\omega^2 + 1)}\,d\omega, \qquad (1.31)$$

where $\tilde{K}_h^*(\omega) = \tilde{K}_h(-\omega)$, $L(\omega) = |\tilde{K}_h(\omega)|^2 = \tilde{K}_h^*(\omega)\tilde{K}_h(\omega)$, and $\tilde{K}_h(\omega)$, $\tilde{u}_\delta(\omega)$ are the Fourier transforms of the functions $K_h(y)$, $u_\delta(x)$; e.g.,

$$\tilde{u}_\delta(\omega) = \int_{-\infty}^{+\infty} u_\delta(x)e^{i\omega x}\,dx.$$

If we substitute the expression for $\tilde{u}_\delta(\omega)$ into (1.31) and change the order of integration, then $z_\eta^\alpha(s)$ takes the form

$$z_\eta^\alpha(s) = \int_{-\infty}^{+\infty} K^\alpha(x - s)u_\delta(x)\, dx, \qquad (1.32)$$

where the inversion kernel $K^\alpha(t)$ has the form

$$K^\alpha(t) = \frac{1}{2\pi} \int_{-\infty}^{+\infty} \frac{\tilde{K}_h^*(\omega)e^{i\omega t}}{L(\omega) + \alpha(\omega^2 + 1)}\, d\omega.$$

Since in the solution of practical problems, $u_\delta(x)$ usually has bounded support, the integral in (1.32) extends only over the domain in which $u_\delta(x)$ is nonzero. Thus, to find $z_\eta^\alpha(s)$ for a fixed α it suffices to numerically find the Fourier transform of the kernel $\tilde{K}_h(\omega)$, then to construct the inversion kernel $K^\alpha(t)$ using standard programs for numerically computing integrals of rapidly oscillating functions, and then use (1.32). The problem of choosing the regularization parameter α has been considered in §2. Consider the case when the solution and the kernel have local supports. In this case equation (1.30) can be written as

$$Az = \int_0^{2a} K(x - s)z(s)\, ds = u(x), \qquad x \in [0, 2a];$$
$$\qquad (1.33)$$
$$A \colon \mathcal{W}_2^1[a, b] \longrightarrow L_2[0, 2a].$$

The operator A (kernel K) can be given both exactly or approximately. Suppose the following conditions on the supports of the functions participating in the equation hold:

$$\operatorname{supp} K(y) \subseteq \left[-\frac{l}{2}, \frac{l}{2}\right], \qquad \operatorname{supp} z(s) \subseteq \left[a - \frac{l_z}{2}, a + \frac{l_z}{2}\right],$$

where $a > 0$, $l > 0$, $l_z \geq 0$, $2l + l_z \leq 2a$ (l_z is the length of the support of the solution $z(s)$).

LEMMA 1.8. *Let* $u(x) \in A\mathcal{W}_2^1[0, 2a]$, $A \neq 0$. *Then the solution of* (1.33) *is unique.*

PROOF. It suffices to prove that the homogeneous equation has only the trivial solution. Assume this to be not true. Then there is a function $z(s) \neq 0$ such that

$$\int_0^{2a} K(x - s)z(s)\, ds = 0.$$

Since $K(y)$ and $z(s)$ have local supports, by defining them to be equal to zero on the remainder of the axis and after taking Fourier transforms, we have

$$\tilde{K}(\omega)\tilde{z}(\omega) = 0,$$

where $\tilde{K}(\omega)$, $\tilde{z}(\omega)$ are the images of $K(y)$, $z(s)$. The functions $K(y)$, $z(s)$ have local supports, so $\tilde{K}(\omega)$ and $\tilde{z}(\omega)$ are analytic functions (see [23]) and hence $z(s) \equiv 0$. □

By the requirements imposed on the supports, after having defined $z(s)$ and $u(x)$ on $[0, 2a]$ and $K(y)$ on $[-a, a]$ we can extend them periodically with period $2a$ onto the real axis. After this, (1.30) can be regarded on the whole real axis.

We now introduce uniform grids for x and s:

$$x_k = s_k = k\Delta x, \qquad \Delta x = \frac{2a}{n}, \quad k = 0, \ldots, n-1$$

(in the sequel n is assumed to be even). For simplicity we approximate (1.30) by the rectangle formula:

$$\sum_{j=0}^{n-1} K(x_k - s_j) z(s_j)\Delta x = u(x_k).$$

Put $u_k = u(x_k)$, $z_j = z(s_j)$, $K_{k-j} = K(x_k - s_j)$, $T = 2a$. We define the *discrete Fourier transform* [17], [125] of functions f_k of a discrete variable k (that are periodic with period n: $f_{n+k} = f_k$, $\forall k$) as follows:

$$\tilde{f}_m = \sum_{k=0}^{n-1} f_k e^{-i\omega_m x_k} = \sum_{k=0}^{n-1} f_k e^{-i2\pi\frac{mk}{n}}, \qquad m = 0, \ldots, n-1.$$

Here, $\omega_m = m\Delta\omega$, $\Delta\omega = 2\pi/T$.

The inverse discrete Fourier transform has the form

$$f_k = \frac{1}{n}\sum_{m=0}^{n-1} \tilde{f}_m e^{i\omega_m x_k}, \qquad k = 0, \ldots, n-1.$$

The proof of this fact follows directly from the relations

$$\sum_{m=0}^{n-1} e^{2\pi i\frac{mk}{n}} = \begin{cases} n & \text{if } k \text{ divides } n, \\ 0 & \text{if } k \text{ does not divide } n. \end{cases}$$

Plancherel's equality also follows directly from these relations. For real f_k we can write it as

$$\sum_{k=0}^{n-1} f_k^2 = \frac{1}{n}\sum_{m=0}^{n-1} |\tilde{f}_m|^2.$$

In exactly the same way we can obtain the analog of the *convolution theorem* for the discrete Fourier transform, to wit

$$\sum_{k=0}^{n-1}\left(\sum_{j=0}^{n-1} K_{k-j} z_j \Delta x\right) e^{-i\omega_m x_k} = \Delta x \sum_{j=0}^{n-1} z_j \sum_{k=0}^{n-1} K_{k-j} e^{-2\pi i\frac{mk}{n}} =$$

$$= \Delta x \sum_{j=0}^{n-1} z_j e^{-2\pi i\frac{mj}{n}} \sum_{p=-j}^{n-1-j} K_p e^{-\pi i\frac{mp}{n}} = \Delta x \tilde{z}_m \tilde{K}_m.$$

The periodicity of K_p with period n is essential here.

We now write down the finite-difference approximation of the functional $M^\alpha[z]$ for equation (1.33):

$$\hat{M}^\alpha[z] = \sum_{k=0}^{n-1} \left(\sum_{j=0}^{n-1} K_{k-j} z_j \Delta x - u_k \right)^2 \Delta x + \alpha \sum_{k=0}^{n-1} z_k^2 \Delta x + \alpha \sum_{k=0}^{n-1} \left(z'(x_k) \right)^2 \Delta x.$$

Here, $z(s)$ and the quantities $\tilde{z}_m$ are related by

$$z(s) = \frac{1}{n} \sum_{m=0}^{n-1} \tilde{z}_m e^{i\omega_m s},$$

where $\tilde{z}_m$ is the discrete Fourier transform of z_k. Then for the discrete Fourier transform coefficients of the vector $z'(x_k)$ we have [133]

$$z'(x_k) = \frac{1}{n} \sum_{m=0}^{n-1} i\omega_m \tilde{z}_m e^{i\omega_m x_k},$$

$$\widetilde{(z'(x_k))}_m = i\omega_m \tilde{z}_m; \qquad k, m = 0, \ldots, n-1.$$

The functional $\hat{M}^\alpha[z]$ approximating $M^\alpha[z]$ can now easily be written as

$$\hat{M}^\alpha[z] = \frac{\Delta x}{n} \sum_{m=0}^{n-1} \left(|\tilde{K}_m|^2 (\Delta x)^2 \tilde{z}_m \tilde{z}_m^* - 2\Delta x \tilde{K}_m \tilde{u}_m \tilde{z}_m^* + |\tilde{u}_m|^2 + \alpha(1 + \omega_m^2) \tilde{z}_m \tilde{z}_m^* \right).$$

Hence the minimum of $\hat{M}^\alpha[z]$ is attained on the vector with discrete Fourier transform coefficients [60]

$$\tilde{z}_m = \frac{\tilde{K}_m^* \tilde{u}_m \Delta x}{|\tilde{K}_m|^2 (\Delta x)^2 + \alpha \left(1 + \left(\frac{\pi}{a} \right)^2 m^2 \right)}, \qquad m = 0, \ldots, n-1. \qquad (1.34)$$

Applying the inverse discrete Fourier transform, we find $z_\eta^\alpha(s)$ at the grid points s_k.

REMARK. We can propose the somewhat different approximating functional

$$M^\alpha[z] = \|A_h z - u_\delta\|_{L_2[0,2a]}^2 + \alpha \|z\|_{\mathcal{W}_2^1[0,2a]}^2,$$

which is based on another approximation of the expression $\Omega_1 = \int_0^{2a} \left(z'(s) \right)^2 ds$.

Approximate Ω_1 by

$$\hat{\Omega}_1 = \Delta x \sum_{k=0}^{n-1} \left(\frac{z_{k+1} - z_k}{\Delta x}\right)^2 = \frac{\Delta x}{n} \frac{1}{(\Delta x)^2} \sum_{k=0}^{n-1} |\tilde{z}_{k+1} - \tilde{z}_k|^2 =$$

$$= \frac{\Delta x}{n} \frac{1}{(\Delta x)^2} \sum_{m=0}^{n-1} \left|\sum_{k=0}^{n-1}(z_{k+1} - z_k)e^{-i\omega_m x_k}\right|^2 =$$

$$= \frac{\Delta x}{n} \frac{1}{(\Delta x)^2} \sum_{m=0}^{n-1} \left|\sum_{p=1}^{n} z_p e^{-i\omega_m(x_p - \Delta x)} - \tilde{z}_m\right|^2 =$$

$$= \frac{\Delta x}{n} \frac{1}{(\Delta x)^2} \sum_{m=0}^{n-1} \left|\tilde{z}_m - e^{i\omega_m \Delta x}\tilde{z}_m\right|^2 = \frac{\Delta x}{n} \frac{1}{(\Delta x)^2} \sum_{m=0}^{n-1} |\tilde{z}_m|^2 |1 - e^{i\omega_m \Delta x}|^2 =$$

$$= \frac{\Delta x}{n} \sum_{m=0}^{n-1} |\tilde{z}_m|^2 \left|\frac{2}{\Delta x}\sin\frac{\omega_m \Delta x}{2}\right|^2 = \frac{\Delta x}{n} \sum_{m=0}^{n-1} |\tilde{z}_m|^2 \left(\frac{2}{\Delta x}\sin\frac{\pi m}{n}\right)^2.$$

When using this approximation in the expressions for the extremals of (1.34), and also in the expressions given below for the functions ρ_η^κ, Φ_η, γ_η, β_η we have to replace the expression $1 + (\pi m/n)^2$ by

$$1 + \left(\frac{2}{\Delta x}\sin\frac{\pi m}{n}\right)^2.$$

The method described above has a simple and fast computer implementation, since there are fast Fourier transform methods [207], [125], [193] for which standard programs have been written [142]. When using the generalized discrepancy principle for choosing the regularization parameter, we need the functions β_η, γ_η; in the case under consideration these can be computed according to the formulas

$$\gamma_\eta(\alpha) = \frac{\Delta x}{n} \sum_{m=0}^{n-1} \frac{|\tilde{K}_m|^2 (\Delta x)^2 |\tilde{u}_m|^2 \left(1 + \frac{\pi m}{a}\right)^2}{\left[|\tilde{K}_m|^2 (\Delta x)^2 + \alpha\left(1 + \left(\frac{\pi m}{a}\right)^2\right)\right]^2},$$

$$\beta_\eta(\alpha) = \frac{\Delta x}{n} \sum_{m=0}^{n-1} \frac{\alpha^2 \left(1 + \left(\frac{\pi m}{a}\right)^2\right) |\tilde{u}_m|^2}{\left[|\tilde{K}_m|^2 (\Delta x)^2 + \alpha\left(1 + \left(\frac{\pi m}{a}\right)^2\right)\right]^2}.$$

The derivative of $\gamma_\eta(\alpha)$ can be readily computed from

$$\gamma_\eta'(\alpha) = -\frac{2\Delta x}{n} \sum_{m=0}^{n-1} \frac{|\tilde{K}_m|^2 (\Delta x)^2 |\tilde{u}_m|^2 \left(1 + \left(\frac{\pi m}{a}\right)^2\right)^2}{\left[|\tilde{K}_m|^2 (\Delta x)^2 + \alpha\left(1 + \left(\frac{\pi m}{a}\right)^2\right)\right]^3}.$$

The derivatives of $\beta_\eta(\alpha)$ and $\rho_\eta^\kappa(\alpha)$ can be computed using (1.21).

As an example of the application of the methods described above we consider the following model problem.

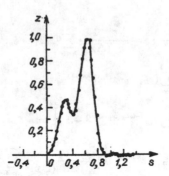

FIGURE 1.4. Model problem for a one-dimensional equation of convolution type.

Suppose we are given the equation

$$\int_0^1 K(x-s)z(s)\,ds = u(x), \qquad x \in [0,2], \quad K(y) = \exp\left\{-80(y-0.5)^2\right\},$$

with local support in $(0,1)$,

$$\overline{z}(s) = \left(\frac{\exp\left\{-\frac{(s-0.3)^2}{0.03}\right\} + \exp\left\{-\frac{(s-0.7)^2}{0.03}\right\}}{0.9550408} - 0.052130913\right) 1.4 \cdot s.$$

The results of the computations of the model equation with error levels $h^2 = 6.73 \times 10^{-9}$, $\delta^2 = 1.56 \times 10^{-6}$ are given in Figure 1.4.

We will now consider two-dimensional integral equations of convolution type:

$$Az = \int_{-\infty}^{+\infty} \int_{-\infty}^{+\infty} K(x-s, y-t)z(s,t)\,ds\,dt = u(x,y), \tag{1.35}$$

$$-\infty < x, y < +\infty.$$

Suppose the kernel $K(v,w)$ and the righthand side $u(x,y)$ belong to $L_2[(-\infty,+\infty) \times (-\infty,+\infty)]$, while the exact solution $\overline{z}(s,t) \in \mathcal{W}_2^2[(-\infty,+\infty) \times (-\infty,+\infty)]$,[1] and the operator A is continuous and single-valued. Suppose that instead of the exact solution $\overline{u}$ and kernel K we know only approximate values u_δ and K_h such that

$$\|u_\delta - \overline{u}\|_{L_2} \le \delta, \qquad \|A - A_h\|_{\mathcal{W}_2^2 \to L_2} \le h.$$

Consider the Tikhonov functional

$$M^\alpha[z] = \|A_h z - u_\delta\|_{L_2}^2 + \alpha\|z\|_{\mathcal{W}_2^2}^2.$$

For each $\alpha > 0$ there is a unique extremal z_η^α of the Tikhonov functional realizing the minimum of $M^\alpha[z]$. When choosing $\alpha = \alpha(\eta)$ in accordance with the generalized discrepancy principle (see §2), $z_\eta^{\alpha(\eta)}$ converges, in the norm of $\mathcal{W}_2^2$ and as $\eta \to 0$, to the

---

[1]The space $\mathcal{W}_2^2$ is the space of functions having generalized derivatives of order two that are square integrable. The imbedding theorem [153] implies that convergence in the norm of $\mathcal{W}_2^2$ implies uniform convergence on every rectangle $[a, b] \times [c, d]$. This determines the choice $Z = \mathcal{W}_2^2$.

exact solution. Hence it also converges uniformly on every rectangle $[a, b] \times [c, d]$ [153]. As for one-dimensional equations, we can readily write down the extremal of $M^\alpha[z]$:

$$z_\eta^{\alpha(\eta)} = \frac{1}{4\pi^2} \int_{-\infty}^{+\infty} \int_{-\infty}^{+\infty} \frac{\tilde{K}_h^*(\omega, \Omega)\tilde{u}_\delta(\omega, \Omega)e^{i\omega s + i\Omega t}}{L(\omega, \Omega) + \alpha\left(1 + (\omega^2 + \Omega^2)^2\right)} \, d\omega \, d\Omega,$$

where $\tilde{K}_h^*(\omega, \Omega) = \tilde{K}_h(-\omega, -\Omega)$, $L(\omega, \Omega) = |\tilde{K}_h(\omega, \Omega)|^2$, and $\tilde{K}_h(\omega, \Omega)$, $\tilde{u}_\delta(\omega, \Omega)$ are the Fourier transforms of $K_h(v, w)$ and $u_\delta(x, y)$, defined by

$$\tilde{u}_\delta(\omega, \Omega) = \int_{-\infty}^{+\infty} \int_{-\infty}^{+\infty} u_\delta(x, y)e^{-i\omega x - i\Omega y} \, dx \, dy,$$

$$\tilde{K}_h(\omega, \Omega) = \int_{-\infty}^{+\infty} \int_{-\infty}^{+\infty} K_h(v, w)e^{-i\omega v - i\Omega w} \, dv \, dw.$$

We may choose $z_\eta^\alpha(s, t)$ in a form similar to (1.32):

$$z_\eta^\alpha(s, t) = \int_{-\infty}^{+\infty} \int_{-\infty}^{+\infty} K_\alpha(x - s, y - t)u_\delta(x, y) \, dx \, dy,$$

where

$$K^\alpha(v, w) = \frac{1}{4\pi^2} \int_{-\infty}^{+\infty} \int_{-\infty}^{+\infty} \frac{\tilde{K}_h^*(\omega, \Omega)e^{-i\omega v - i\Omega w}}{L(\omega, \Omega) + \alpha\left(1 + (\omega^2 + \Omega^2)^2\right)} \, d\omega \, d\Omega.$$

We will now consider the case when the exact solution $\overline{z}(s, t)$ and the kernel $K(v, w)$ have local supports:

$$\text{supp}\, K(v, w) \subseteq [l_1, L_1] \times [l_2, L_2],$$

$$\text{supp}\, \overline{z}(s, t) \subseteq [a, A] \times [b, B].$$

Then $\overline{u}(x, y)$ has supp $\overline{u}(x, y) \subseteq [c, C] \times [d, D]$ where $c = a + l_1$, $C = A + L_1$, $d = b + l_2$, $D = B + L_2$. We will assume that the approximate righthand side $u_\delta(x, y)$ has local support in $[c, C] \times [d, D]$ and that the approximate solution has local support in $[a, A] \times [b, B]$.

By considerations similar to the one-dimensional case given above, we arrive at

$$\int_0^{2R} \int_0^{2r} K(x - s, y - t)z(s, t) \, ds \, dt = u(x, y), \qquad (1.36)$$

in which the local supports of $z(s, t)$ and $u(x, y)$ lie in the interior of the rectangle $[0, 2r] \times [0, 2R]$, while outside their supports z and u are put equal to zero on the whole rectangle.

After performing a similar procedure for $K(v, w)$, we will assume that all functions are defined periodically (with period $2r$ in the first and $2R$ in the second argument), and consider equation (1.36) on the whole plane.

Introducing uniform grids in (x, y) and (s, t):

$$x_k = s_k = k\Delta x, \qquad y_l = t_l = l\Delta y,$$

$$\Delta x = \frac{2r}{n_1}, \qquad \Delta y = \frac{2R}{n_2}, \qquad k = 0, \ldots, n_1 - 1, \quad l = 0, \ldots, n_2 - 1$$

(n_1 and n_2 are assumed to be even) and approximating the equation by the rectangle formula, we obtain

$$\sum_{m=0}^{n_1-1}\sum_{j=0}^{n_2-1} K_{k-m,l-j}z_{mj}\Delta x\Delta y = u_{kl},$$

where

$$u_{kl} = u(x_k,y_l), \qquad z_{mj} = z(s_m,t_j), \qquad K_{k-m,l-j} = K_h(x_k - s_m, y_l - t_j).$$

The discrete Fourier transform is defined as

$$\tilde{f}_{mn} = \sum_{k=0}^{n_1-1}\sum_{l=0}^{n_2-1} f_{kl}e^{-i\omega_m x_k - i\Omega_n y_l},$$

$$\omega_m = m\Delta\omega, \quad \Delta\omega = \frac{\pi}{r}, \quad \Omega_n = n\Delta\Omega, \quad \Delta\Omega = \frac{\pi}{R},$$

$$m = 0,\ldots,n_1-1, \qquad n = 0,\ldots,n_2-1,$$

and the inverse transform takes the form

$$f_{kl} = \frac{1}{n_1 n_2}\sum_{m=0}^{n_1-1}\sum_{n=0}^{n_2-1}\tilde{f}_{mn}e^{i\omega_m x_k + i\Omega_n y_l},$$

$$k = 0,\ldots,n_1-1, \qquad l = 0,\ldots,n_2-1.$$

The two-dimensional analog of the Plancherel equality has the form

$$\sum_{k=0}^{n_1-1}\sum_{l=0}^{n_2-1} |f_{kl}|^2 = \frac{1}{n_1 n_2}\sum_{m=0}^{n_1-1}\sum_{n=0}^{n_2-1} |\tilde{f}_{mn}|^2,$$

and the convolution theorem is:

$$\sum_{k=0}^{n_1-1}\sum_{l=0}^{n_2-1}\left\{\sum_{p=0}^{n_1-1}\sum_{j=0}^{n_2-1} K_{k-p,l-j}z_{pj}\Delta x\Delta y\right\}e^{-i\omega_m x_k - i\Omega_n y_l} = \tilde{K}_{mn}\tilde{z}_{mn}\Delta x\Delta y,$$

$$m = 0,\ldots,n_1-1, \qquad n = 0,\ldots,n_2-1.$$

We write down the finite-difference approximation of the functional $M^\alpha[z]$ for equation (1.35):

$$\hat{M}^\alpha[z] = \sum_{k=0}^{n_1-1}\sum_{l=0}^{n_2-1}\left(\sum_{p=0}^{n_1-1}\sum_{j=0}^{n_2-1} K_{k-p,l-j}z_{pj}\Delta x\Delta y - u_{kl}\right)^2 \Delta x\Delta y +$$

$$+ \alpha\sum_{k=0}^{n_1-1}\sum_{l=0}^{n_2-1}\left\{z_{kl}^2 + \left[\frac{\partial^2 z(s_k,t_l)}{\partial s^2}\right]^2 + 2\left[\frac{\partial^2 z(s_k,t_l)}{\partial s\partial t}\right]^2 + \left[\frac{\partial^2 z(s_k,t_l)}{\partial t^2}\right]^2\right\}\Delta x\Delta y.$$

As in the previous section we arrive at

$$\left(\frac{\widetilde{\partial^2 z(s_k, t_l)}}{\partial s^2}\right)_{mn} = -\omega_m^2 \tilde{z}_{mn},$$

$$\left(\frac{\widetilde{\partial^2 z(s_k, t_l)}}{\partial s \partial t}\right)_{mn} = -\omega_m \Omega_n \tilde{z}_{mn},$$

$$\left(\frac{\widetilde{\partial^2 z(s_k, t_l)}}{\partial t^2}\right)_{mn} = -\Omega_n^2 \tilde{z}_{mn},$$

$$k, m = 0, \dots, n_1 - 1, \qquad l, n = 0, \dots, n_2 - 1,$$

while the functional $\hat{M}^\alpha[z]$ can be written as

$$\hat{M}^\alpha[z] = \frac{\Delta x \Delta y}{n_1 n_2} \sum_{m=0}^{n_1-1} \sum_{n=0}^{n_2-1} \left\{ |\tilde{K}_{mn} \tilde{z}_{mn} \Delta x \Delta y - \tilde{u}_{mn}|^2 + \right. \tag{1.37}$$

$$\left. + \alpha \left[1 + (\omega_m^2 + \Omega_n^2)^2\right] |\tilde{z}_{mn}|^2 \right\}.$$

The minimum of the functional (1.37) is attained on the vector with Fourier coefficients

$$z_{mn}^\alpha = \frac{\tilde{K}_{mn}^* \tilde{u}_{mn} \Delta x \Delta y}{|\tilde{K}_{mn}|^2 (\Delta x \Delta y)^2 + \alpha \left[1 + (\omega_m^2 + \Omega_n^2)^2\right]},$$

$$m = 0, \dots, n_1 - 1, \qquad n = 0, \dots, n_2 - 1.$$

On the grid (s_k, t_l) the solution $z_\eta^\alpha(s, t)$ can be obtained by inverse Fourier transformation:

$$z_\eta^\alpha(s_k, t_l) = \frac{1}{n_1 n_2} \sum_{m=0}^{n_1-1} \sum_{n=0}^{n_2-1} z_{mn}^\alpha e^{i\omega_m s_k + i\Omega_n t_l}. \tag{1.38}$$

As for one-dimensional equations, the expressions for $\beta_\eta(\alpha)$, $\gamma_\eta(\alpha)$, $\rho_\eta(\alpha)$ and their derivatives with respect to α can be obtained without computing the extremal z_η^α. We give some of these formulas:

$$\beta_\eta(\alpha) = \frac{\Delta x \Delta y}{n_1 n_2} \sum_{m=0}^{n-1-1} \sum_{n=0}^{n_2-2} \frac{\alpha^2 |\tilde{u}_{mn}|^2 \left(1 + (\omega_m^2 + \Omega_n^2)^2\right)}{\left\{|\tilde{K}_{mn}|^2 (\Delta x \Delta y)^2 + \alpha \left[1 + (\omega_m^2 + \Omega_n^2)^2\right]\right\}^2},$$

$$\gamma_\eta(\alpha) = \|z_\eta^\alpha\|_{W_2^2}^2 = \frac{\Delta x \Delta y}{n_1 n_2} \sum_{m=0}^{n-1-1} \sum_{n=0}^{n_2-2} \frac{|\tilde{K}_{mn}|^2 |\tilde{u}_{mn}|^2 (\Delta x \Delta y)^2 \left[1 + (\omega_m^2 + \Omega_n^2)^2\right]}{\left\{|\tilde{K}_{mn}|^2 (\Delta x \Delta y)^2 + \alpha \left[1 + (\omega_m^2 + \Omega_n^2)^2\right]\right\}^2}.$$

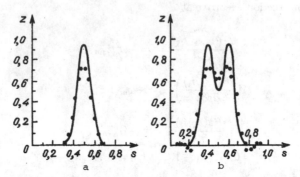

FIGURE 1.5. Model problem for the two-dimensional integral equation of convolution type: a) with section $s = 0.28125$; b) with section $s = 0.46875$.

The derivative of $\rho_\eta(\alpha) = \beta_\eta(\alpha) - \left(\delta + h\sqrt{\gamma_\eta(\alpha)}\right)^2$ is:

$$\rho'_\eta(\alpha) = \frac{\Delta x \Delta y}{n_1 n_2} \sum_{m=0}^{n_1-1} \sum_{n=0}^{n_2-1} \frac{|\tilde{K}_{mn}|^2 |\tilde{u}_{mn}|^2 (\Delta x \Delta y)^2 \left[1 + (\omega_m^2 + \Omega_n^2)^2\right]}{\left\{|\tilde{K}_{mn}|^2 (\Delta x \Delta y)^2 + \alpha \left[1 + (\omega_m^2 + \Omega_n^2)^2\right]\right\}^3} \times$$

$$\times \left\{\alpha + \left(\delta + h\sqrt{\gamma_\eta(\alpha)}\right) \frac{h\left[1 + (\omega_m^2 + \Omega_n^2)^2\right]}{\sqrt{\gamma_\eta(\alpha)}}\right\}.$$

In the appendices to this book we give a standard program for solving a two-dimensional integral equation involving a difference of the arguments. This program is based on the technique of the fast Fourier transform with choice of the parameter α in accordance with the generalized discrepancy principle.

We give as example a test calculation using the equation

$$\int_0^1 \int_0^1 K(x - s, y - t) z(s, t) \, ds \, dt = u(x, y)$$

with kernel

$$K(v, w) = \begin{cases} \exp\left\{-20\left[(v - 0.5)^2 + (y - 0.5)^2\right]\right\} & \text{if } 0 \leq v, w \leq 1, \\ 0 & \text{otherwise.} \end{cases}$$

We take the exact solution to be

$$\bar{z}(s, t) = \left(\frac{\exp\left\{-\frac{(s-0.3)^2}{0.03}\right\} + \exp\left\{-\frac{(s-0.7)^2}{0.03}\right\}}{0.955040800} - 0.052130913\right) \exp\left\{-\frac{(t - 0.5)^2}{0.03}\right\}.$$

We are given grid dimensions $n_1 = n_2 = 32$, and error levels $\delta^2 = 6.3 \times 10^{-7}$, $h^2 = 3.7 \times 10^{-11}$. The results are given in Figure 1.5.

8. Nonlinear ill-posed problems

When solving nonlinear ill-posed problems a number of difficulties come about which are not only of a numerical nature. The main assertions and theorems in §1 use the linearity of the operator A in an essential manner. For nonlinear operators similar results can be obtained by using the scheme of compact imbedding.

A detailed investigation of methods for solving nonlinear ill-posed problems is beyond the scope of this book, and here we will restrict ourselves to the description of two approaches to the solution of nonlinear ill-posed problems.

As before, we will consider an equation

$$Az = u, \qquad z \in Z, \quad u \in U.$$

The spaces Z and U are assumed to be equipped with a norm, and the operator $A \colon Z \to U$ is taken continuous and bijective. We will also assume that the perturbed operator A_h is continuous. Moreover, we assume that $\|Az - A_h z\| \le \psi(h, \|z\|)$, where $\psi(h, y)$ is a function that is continuous in the set of arguments for $h \ge 0$, $y \ge 0$, is monotonically nondecreasing in the first argument, nonnegative, and such that $\psi(h, y) \Rightarrow 0$ as $h \to 0$ uniformly with respect to y on any interval $[0, C]$.

Let V be a bijective operator acting from the Hilbert space X into Z. Suppose V is *compact*, i.e. transforms weakly convergent sequences into strongly convergent sequences. For example, if X is compactly imbedded in Z, then V could be the imbedding operator. Let D be the closed convex set of constraints of the problem, $D \subseteq X$.

We will assume that $\overline{z} \in VD \subseteq Z$. As before, $\overline{u} = A\overline{z}$, $\|\overline{u} - u_\delta\| \le \delta$. In this way we arrive at the problem

$$AVx = u, \qquad AV \colon X \to U, \qquad x \in D; \tag{1.39}$$

and instead of $\overline{u}$, A we are given approximations of them. The operator AV of the new problem is compact and bijective [29].

Consider the extremal problem
find

$$\inf_{x \in X_\eta} \|x\|, \qquad X_\eta = \left\{ x \colon x \in D, \ \|A_h Vx - u_\delta\| \le \delta + \psi(h, \|Vx\|) \right\}. \tag{1.40}$$

It is obvious that the set X_η is not empty, since it contains at least a point $\overline{x}$ such that $\overline{z} = V\overline{x}$. Consequently, this problem is equivalent to the problem:
find

$$\inf_{x \in X_\eta \cap \overline{S}(0, R)} \|x\|, \tag{1.41}$$

where $\overline{S}(0, R)$ is the closed ball in X with center at zero and radius $R = \|\overline{x}\|$.

To prove the solvability of this problem it suffices to show that $X_\eta \cap \overline{S}(0, R)$ is weakly compact in X, and to use further the fact that the convex continuous functional $f(x) = \|x\|$ in the Hilbert space X is weakly lower semicontinuous, and then apply the Weierstrass theorem [33].

LEMMA 1.9. *The set $X_\eta \cap \overline{S}(0, R)$ is weakly compact in X.*

PROOF. The set $D \cap \overline{S}(0, R)$ is nonempty and bounded, and X is a Hilbert space, so $D \cap \overline{S}(0, R)$, and hence also $X_\eta \cap \overline{S}(0, R)$, is relatively weakly compact. We show that $X_\eta \cap \overline{S}(0, R)$ is weakly closed. Let $\{x_n\}$, $x_n \in X_\eta \cap \overline{S}(0, R)$, be a sequence weakly converging to $x^* \in X$. Since $X_\eta \cap \overline{S}(0, R) \subseteq D \cap \overline{S}(0, R)$ and $D \cap \overline{S}(0, R)$ is convex and closed, we see that $x^* \in D \cap \overline{S}(0, R)$. Using the relations

$$\|A_h V x^* - u_\delta\| = \|A_h V x^* - A_h V x_n + A_h V x_n - u_\delta\| \leq$$
$$\leq \delta + \psi(h, \|V x_n\|) + \|A_h V x^* - A_h V x_n\|,$$

the compactness of the operators V and AV and the continuity of $\psi(h, y)$ with respect to y, we obtain after limit transition:

$$\|A_h V x^* - u_\delta\| \leq \delta + \psi(h, \|V x^*\|),$$

i.e. $x^* \in X_\eta \cap \overline{S}(0, R)$. $\square$

REMARK. If $0 \in D$ ·but $0 \notin X_\eta$, i.e. $\|A_h V 0 - u_\delta\| > \delta + \psi(h, \|V 0\|)$, then the problem (1.40) is equivalent to the problem
    find
$$\inf \|x\|, \qquad x \in \left\{ x : x \in D, \|A_h V x - u_\delta\| = \delta + \psi(h, \|V x\|) \right\}.$$

In fact, assume that these problems are not equivalent. Then we can find a solution $x_\eta \in D$ of (1.40) satisfying the inequality $\|A_h V x_\eta - u_\delta\| < \delta + \psi(h, \|V x_\eta\|)$. The function $\Phi(\lambda) = \|A_h V \lambda x_\eta - u_\delta\| - \delta - \psi(h, \|V \lambda x_\eta\|)$ is continuous and $\Phi(0) > 0$ while $\Phi(1) < 0$. Therefore there is a $\lambda^* \in (0, 1)$ at which $\Phi(\lambda^*) = 0$. However, $\|\lambda^* x_\eta\| = \lambda^* \|x_\eta\| < \|x_\eta\|$, contradicting the fact that x_η is a solution of (1.40).

So, for arbitrary $h \geq 0$, $\delta \geq 0$, $u_\delta \in U$ such that $\|A V \overline{x} - u_\delta\| \leq \delta$, the problem (1.40) is solvable. Let X_η^* be the set of solutions of this problem. Let $\eta_n \to 0$ be a given sequence.

THEOREM 1.6. *A sequence $\{x_n\}$ of arbitrary elements from the sets $X_{\eta_n}^*$ converges to an $\overline{x} \in D$ in the norm of X.*

PROOF. For any η_n we have $\|x_n\| \leq \|\overline{x}\|$. Since X is a Hilbert space we can extract a subsequence $\{x_{n_k}\}$ from $\{x_n\}$ that converges weakly to some $x^* \in D \cap \overline{S}(0, R)$. Moreover,

$$\|A V x_{n_k} - \overline{u}\| \leq \|A_{h_{n_k}} V x_{n_k} - u_{\delta_{n_k}}\| + \|A V x_{n_k} - A_{h_{n_k}} V x_{n_k}\| + \|u_{\delta_{n_k}} - \overline{u}\| \leq$$
$$\leq 2 \left(\delta_{n_k} + \psi(h_{n_k}, \|V x_{n_k}\|) \right).$$

The sequence $\{x_{n_k}\}$ converges weakly, therefore the sequence $\|V x_{n_k}\|$ is bounded: $0 \leq \|V x_{n_k}\| \leq C$. Transition to the limit as $k \to \infty$, the use of the properties of the function $\psi(h, y)$ and the compactness of AV, give

$$\|A V x^* - \overline{u}\| = 0.$$

Since AV is bijective, we hence have $x^* = \overline{x}$. Further, since $\|x\|$ is weakly lower semicontinuous,

$$\|x^*\| \leq \liminf_{k \to \infty} \|x_{n_k}\| \leq \limsup_{k \to \infty} \|x_{n_k}\| \leq \|\overline{x}\|,$$

i.e.

$$\lim_{k\to\infty} \|x_{n_k}\| = \|\overline{x}\|.$$

Since X is a Hilbert space, $x_{n_k} \to \overline{x}$ in the norm of X. This is true for any sequence $\{x_{n_k}\}$, so $\lim_{n\to\infty} x_n = \overline{x}$. $\square$

REMARKS. 1. Since V is compact, $z_n = Vx_n$ converges to $\overline{z}$ (the solution of (1.1)) in the norm of Z.

2. Suppose AV is not bijective. Let $\overline{X}$ be the set of solutions of the equation $AVx = \overline{u}$, $\overline{X} \subseteq D$. Applying Lemma 1.9 with $h = 0$, $\delta = 0$, $X_\eta = \overline{X}$, we see that this equation has the normal solution $\overline{x}$. The proof of Theorem 1.6 implies that the algorithm for searching an approximate solution as a solution of the extremal problem (1.40) guarantees that the sequence of regularized approximations converges to the normal solution.

Use of the generalized discrepancy method in solving the problem (1.40) was first proposed in [58] for the case $D = X = W_2^1[a, b]$, $Z = L_2[a, b]$, $U = L_2[a, b]$, V the imbedding operator, and $\psi(h, y) = hy$. In [204] the given problem was investigated under the condition that X is a reflexive space.

Note that in the statement of the problem given above, when dropping the requirement that $\|u_\delta - \overline{u}\| \le \delta$, $\overline{u} \in AVD$, the problem (1.40) can, in general, be unsolvable since the set X_η can be empty.

To construct approximate solutions of ill-posed problems in the case of a nonlinear operator we can use, as in §1, the functional M^α. The first regularization method for solving nonlinear problems using the smoothing functional was given in [167], [168], [182]. We will consider the problem of choosing the regularization parameter in nonlinear problems under the condition that the operator A is exactly known. Moreover, we will consider a concrete regularizer. For other means of specifying regularizers in the smoothing functional when solving nonlinear ill-posed problems, see [178].

Below we will show that this problem is nontrivial, using an example that will show that even when constraints are absent, in nonlinear problems the usual behavior of functions such as the discrepancy (continuity, strong monotonicity) does not take place [62], [64].

We take $Z = U = \mathbf{R}$, the space of real numbers with the natural norm $\|z\|_{\mathbf{R}} = |z|$. Let α_0 be a positive numerical parameter and let $\overline{z}$, $\overline{u}$ fixed elements from $\mathbf{R}$ related by $\overline{z} = \overline{u}/\sqrt{\alpha_0}$, $\overline{u} > 0$. Consider the nonlinear operator $A: Z \to U$ given by

$$Az = \begin{cases} 2\overline{u}, & z \le 0, \\ (\overline{u}^2 - \alpha_0 z^2)^{1/2}, & 0 < z \le \overline{z}, \\ (\alpha_0 z^2 - \overline{u}^2)^{1/2}, & z > \overline{z}. \end{cases}$$

It can be readily seen that a unique solution $\overline{z} \in \mathbf{R}$ corresponds to $u = \overline{u}$. Let δ be a number satisfying the condition $0 < \delta \le \delta_0 < \|\overline{u}\|_{\mathbf{R}}$. We will assume that to each value of the error δ there corresponds an element $u_\delta = \overline{u} \in U$. Consider the functional

$$M^\alpha[z] = \|Az - u_\delta\|_{\mathbf{R}}^2 + \alpha\|z\|_{\mathbf{R}}^2.$$

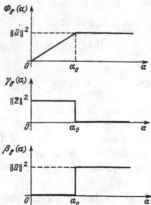

FIGURE 1.6. The graphs of the functions $\Phi_\delta(\alpha)$, $\gamma_\delta(\alpha)$, $\beta_\delta(\alpha)$ for a nonlinear problem.

It is not difficult to see that in our case an extremal of $M^\alpha[z]$ is defined by

$$z_\delta^\alpha = \begin{cases} \overline{z}, & 0 < \alpha < \alpha_0, \\ z^*, & \alpha = \alpha_0, \\ 0, & \alpha_0 < \alpha. \end{cases}$$

Here, z^* is any number in $[0, \overline{z}]$. For the discrepancy we find

$$\beta_\delta(\alpha) = \|Az_\delta^\alpha - u_\delta\|_{\mathbf{R}}^2 = \begin{cases} 0, & 0 < \alpha < \alpha_0, \\ \|\overline{u}\|^2, & \alpha_0 < \alpha. \end{cases}$$

This implies that choosing the regularization parameter in accordance with the discrepancy principle, from the equation

$$\beta_\delta(\alpha) = \|Az_\delta^\alpha - u_\delta\|_{\mathbf{R}}^2 = \delta^2,$$

is impossible, whatever $\delta > 0$ in the interval $(0, \delta_0)$.

This simple example shows that for nonlinear problems the discrepancy principle in the form $\beta_\delta(\alpha) = \delta^2$ is not applicable. In Figure 1.6 we have drawn the graphs of the functions $\Phi_\delta(\alpha) = M^\alpha[z_\delta^\alpha]$, $\gamma_\delta(\alpha) = \|z_\delta^\alpha\|^2$ and $\beta_\delta(\alpha)$ for the given example.

So, assume we are given the problem (1.39) with operators A, V and spaces Z, X, U satisfying the conditions listed above. In addition, regarding $D \subseteq X$ we assume that $0 \in D$, and we assume A to be exactly specified.

Consider the functional

$$M^\alpha[x] = \|AVx - u_\delta\|^2 + \alpha\|x\|^2.$$

The incompatibility measure of (1.39) on D is defined as:

$$\mu_\delta(u_\delta, A) = \inf_{x \in D} \|AVx - u_\delta\|.$$

As in §1, we assume that we know an approximation $\mu_\delta^\kappa(u_\delta, A)$ of μ_δ with accuracy $\kappa \geq 0$. We assume that κ matches δ in the sense that $\kappa(\delta) \to 0$ as $\delta \to 0$.

LEMMA 1.10. *For any $u_\delta \in U$ and any $\alpha > 0$ the problem find*

$$\inf_{x \in D} M^\alpha[x]$$

is solvable, i.e. the set of extremals $X_\delta^\alpha \neq \emptyset$.

The proof of this lemma is in fact given in [168]. When considering problems with constraints we have to take into account that the set $D \cap \overline{S}(0, R)$ is, for any $R > 0$, weakly compact, since X is a Hilbert space and D is convex and closed, and we have to use the continuity of the operator AV, which acts from the weak topology on X into the strong topology of U, and also the weak lower semicontinuity of the functional $\|x\|^2$.

Let Q be a choice operator, mapping the set of extremals X_δ^α to a fixed element $x_\delta^\alpha \in X_\delta^\alpha$: $Q X_\delta^\alpha = x_\delta^\alpha \in X_\delta^\alpha$. As in §4 we consider the auxiliary functions

$$\Phi_\delta(\alpha) = \|AV x_\delta^\alpha - u_\delta\|^2 + \alpha \|x_\delta^\alpha\|^2,$$

$$\gamma_\delta(\alpha) = \|x_\delta^\alpha\|^2, \qquad \beta_\delta(\alpha) = \|AV x_\delta^\alpha - u_\delta\|^2.$$

LEMMA 1.11. *The functions Φ_δ, γ_δ, β_δ have the following properties:*

1. *$\Phi_\delta(\alpha)$ is continuous and concave for $\alpha > 0$.*
2. *$\gamma_\delta(\alpha)$ is monotonically nonincreasing and $\Phi_\delta(\alpha)$, $\beta_\delta(\alpha)$ are monotonically nondecreasing for $\alpha > 0$; moreover, on an interval $(0, \alpha_0)$ on which $x_\delta^\alpha \neq 0$ the function $\Phi_\delta(\alpha)$ is strictly monotone for any $\alpha \in (0, \alpha_0)$.*
3. *The following relations hold:*

$$\lim_{\alpha \to +\infty} \gamma_\delta(\alpha) = \lim_{\alpha \to +\infty} \alpha \gamma_\delta(\alpha) = 0;$$

$$\lim_{\alpha \to +\infty} \Phi_\delta(\alpha) = \lim_{\alpha \to +\infty} \beta_\delta(\alpha) = \|AV0 - u_\delta\|^2 \equiv \nu_\delta^2;$$

$$\lim_{\alpha \downarrow 0} \alpha \gamma_\delta(\alpha) = 0;$$

$$\lim_{\alpha \downarrow 0} \Phi_\delta(\alpha) = \lim_{\alpha \downarrow 0} \beta_\delta(\alpha) = \left(\mu_\delta(u_\delta, A)\right)^2.$$

The proof is similar to the proofs of the corresponding items in Lemma 1.3, §4. Note that if $\overline{x} \in D$, $\|AV\overline{x} - u_\delta\| \leq \delta$, then $\mu_\delta(u_\delta, A) \leq \delta$.

We will assume that $\delta \in (0, \delta_0]$ (δ_0 fixed) and that the natural condition $\nu_\delta \equiv \|AV0 - u_\delta\| > \delta_0 + \mu_\delta^\kappa(u_\delta, A)$ holds. Let C be a number not depending on δ and satisfying the condition

$$1 < C < \frac{\|AV0 - u_\delta\|}{\delta_0 + \mu_\delta^\kappa(u_\delta, A)} = \frac{\nu_\delta}{\delta_0 + \mu_\delta^\kappa(u_\delta, A)}. \tag{1.42}$$

For given C and u_δ we introduce the sets [64]

$$A_\delta^1 = \left\{ \alpha : \alpha > 0,\ \beta_\delta(\alpha) \geq \left(\delta + \mu_\delta^\kappa(u_\delta, A) \right)^2 \right\},$$

$$A_\delta^2 = \left\{ \alpha : \alpha > 0,\ \beta_\delta(\alpha) \leq C^2 \left(\delta + \mu_\delta^\kappa(u_\delta, A) \right)^2 \right\},$$

$$B_\delta^1 = \left\{ \alpha : \alpha > 0,\ \beta_\delta(\alpha) \leq \left(\delta + \mu_\delta^\kappa(u_\delta, A) \right)^2 \right\},$$

$$B_\delta^2 = \left\{ \alpha : \alpha > 0,\ \beta_\delta(\alpha) \geq C^2 \left(\delta + \mu_\delta^\kappa(u_\delta, A) \right)^2 \right\}.$$

LEMMA 1.12. *The sets A_δ^1, A_δ^2, B_δ^1, B_δ^2 are nonempty.*

PROOF. The following assertions hold:

$$A_\delta^1 \neq \emptyset \quad \text{since} \quad \lim_{\alpha \to \infty} \beta_\delta(\alpha) = \nu_\delta^2 > C^2 \left(\delta + \mu_\delta^\kappa(u_\delta, A) \right)^2,$$

$$A_\delta^2 \neq \emptyset \quad \text{since} \quad \lim_{\alpha \downarrow 0} \beta_\delta(\alpha) = \left(\mu_\delta(u_\delta, A) \right)^2 < C^2 \left(\delta + \mu_\delta^\kappa(u_\delta, A) \right)^2,$$

$$B_\delta^1 \neq \emptyset \quad \text{since} \quad \lim_{\alpha \downarrow 0} \beta_\delta(\alpha) = \left(\mu_\delta(u_\delta, A) \right)^2 < \left(\delta + \mu_\delta^\kappa(u_\delta, A) \right)^2,$$

$$B_\delta^2 \neq \emptyset \quad \text{since} \quad \lim_{\alpha \to \infty} \Phi_\delta(\alpha) = \nu_\delta^2 > C^2 \left(\delta + \mu_\delta^\kappa(u_\delta, A) \right)^2.$$

All the limit relations above have been proved in Lemma 1.11. □

LEMMA 1.13. *The set $E_\delta = (A_\delta^1 \cap A_\delta^2) \cup (B_\delta^1 \cap B_\delta^2)$ is nonempty.*

PROOF. Let $A_\delta^1 \cap A_\delta^2 = \emptyset$. Then we cannot choose α in accordance with the discrepancy principle, from the inequality $\left(\delta + \mu_\delta^\kappa(u_\delta, A) \right)^2 \leq \beta_\delta(\alpha) \leq C^2 \left(\delta + \mu_\delta^\kappa(u_\delta, A) \right)^2$. We show that in this case $B_\delta^1 \cap B_\delta^2 \neq \emptyset$. To this end we note that since $\beta_\delta(\alpha)$ is monotonically nondecreasing for $\alpha > 0$, there is an

$$\alpha^* = \sup \left\{ \alpha : \alpha > 0,\ \beta_\delta(\alpha) \leq \left(\delta + \mu_\delta^\kappa(u_\delta, A) \right)^2 \right\} = \sup B_\delta^1,$$

and $\alpha^* > 0$. Moreover, by assumption,

$$\limsup_{\alpha \uparrow \alpha^*} \beta_\delta(\alpha) \leq \left(\delta + \mu_\delta^\kappa(u_\delta, A) \right)^2,$$

$$\liminf_{\alpha \downarrow \alpha^*} \beta_\delta(\alpha) \geq C^2 \left(\delta + \mu_\delta^\kappa(u_\delta, A) \right)^2.$$

But now, by the continuity of $\Phi_\delta(\alpha)$ and the limit relations

$$\lim_{\alpha \downarrow \alpha^*} \Phi_\delta(\alpha) \geq C^2 \left(\delta + \mu_\delta^\kappa(u_\delta, A) \right)^2 + \alpha^* \| x_\delta^{\alpha^*} \|^2 > C^2 \left(\delta + \mu_\delta^\kappa(u_\delta, A) \right)^2$$

(the last inequality is true if $x_\delta^{\alpha^*} \neq 0$),

$$\lim_{\alpha \downarrow \alpha^*} \Phi_\delta(\alpha) = \left(\mu_\delta(u_\delta, A) \right)^2 < C^2 \left(\delta + \mu_\delta^\kappa(u_\delta, A) \right)^2,$$

we can verify that $B_\delta^1 \cap B_\delta^2 \neq \emptyset$.

However, if $x_\delta^{\alpha^*} = 0$, then $X_\delta^\alpha = \{0\}$ and $\Phi_\delta(\alpha) = \beta_\delta(\alpha) = \nu_\delta^2$ for all $\alpha > \alpha^*$, and hence

$$\lim_{\alpha \downarrow \alpha^*} \Phi_\delta(\alpha) = \Phi_\delta(\alpha^*) = \nu_\delta^2 > C^2 \left(\delta + \mu_\delta^\kappa(u_\delta, A)\right)^2. \qquad \square$$

We can now state the *alternative principle for choosing the regularization parameter*: Let $\overline{u} = AV\overline{x}$, $\overline{x} \in D$, $\|\overline{u} - u_\delta\| \le \delta \to 0$ as $\delta \to 0$ and suppose (1.42) is satisfied. Then we can take $x_\delta = x_\delta^{\tilde{\alpha}} \in X_\delta^{\tilde{\alpha}}$ as approximate solution of (1.39), where $\tilde{\alpha} \in E_\delta$.

THEOREM 1.7. *Let* $\delta_n, \kappa_n \to 0$ *as* $n \to \infty$. *Then* $x_{\delta_n} \longrightarrow^{wk} \overline{x}$, *and* $Vx_{\delta_n} \longrightarrow^{str} \overline{z} = V\overline{x}$ *in the norm of* Z.

PROOF. If $\tilde{\alpha}_n \in A_{\delta_n}^1 \cap A_{\delta_n}^2$, by the extremal properties of x_{δ_n} we have

$$\left(\delta_n + \mu_{\delta_n}^{\kappa_n}(u_\delta, A)\right)^2 + \tilde{\alpha}_n \|x_{\delta_n}\|^2 \le \|AVx_{\delta_n} - u_{\delta_n}\|^2 + \tilde{\alpha}_n \|x_{\delta_n}\|^2 \le$$
$$\le \|AV\overline{x} - u_{\delta_n}\|^2 + \tilde{\alpha}_n \|\overline{x}\|^2 \le$$
$$\le \delta_n^2 + \tilde{\alpha}_n \|\overline{x}\|^2 \le \left(\delta + \mu_{\delta_n}^{\kappa_n}(u_{\delta_n}, A)\right)^2 + \tilde{\alpha}_n \|\overline{x}\|^2.$$

Consequently,

$$\|x_{\delta_n}\| \le \|\overline{x}\|.$$

If $\tilde{\alpha} \in B_{\delta_n}^1 \cap B_{\delta_n}^2$, then

$$\Phi_{\delta_n}(\tilde{\alpha}_n) \ge C^2 \left(\delta_n + \mu_{\delta_n}^{\kappa_n}(u_{\delta_n}, A)\right)^2,$$

$$\Phi_{\delta_n}(\tilde{\alpha}_n) \le \|AV\overline{x} - u_{\delta_n}\|^2 + \tilde{\alpha}_n \|\overline{x}\|^2 \le \left(\delta + \mu_{\delta_n}^{\kappa_n}(u_{\delta_n}, A)\right)^2 + \tilde{\alpha}_n \|\overline{x}\|^2.$$

These inequalities imply

$$\frac{C^2 - 1}{C^2} \tilde{\alpha}_n \|x_{\delta_n}\|^2 \le \frac{C^2 - 1}{C^2} \Phi_{\delta_n}(\tilde{\alpha}_n) \le \tilde{\alpha}_n \|\overline{x}\|^2,$$

so that

$$\|x_{\delta_n}\|^2 \le \frac{C^2}{C^2 - 1} \|\overline{x}\|^2.$$

Thus, the sequence $\{x_{\delta_n}\}$ is relatively weakly compact. Further, since in both cases

$$\beta_{\delta_n}(\tilde{\alpha}_n) = \|AVx_{\delta_n} - u_{\delta_n}\|^2 \le C^2 \left(\delta_n + \mu_{\delta_n}^{\kappa_n}(u_{\delta_n}, A)\right)^2 \to 0$$

as $n \to \infty$ and the operator AV is compact, as in Theorem 1.1 we find $x_{\delta_n} \longrightarrow^{wk} \overline{x}$. Consequently, $Vx_{\delta_n} \to V\overline{x} = \overline{z}$ in the norm of Z. $\square$

REMARKS. 1. Thus, $z_{\delta_n} = VX_{\delta_n}$ can be regarded as an approximate solution of equation (1.1) with nonlinear operator A.

2. The alternative discrepancy principle means that if we cannot choose α in accordance with the discrepancy principle $\beta_\delta(\alpha) = (\delta + \mu_\delta^\kappa(u_\delta, A))^2$ (since at the corresponding point the discrepancy function $\beta_\delta(\alpha)$ has a discontinuity), then we have to choose as regularization parameter the left of the discontinuity point in accordance with the principle of the smoothing functional [120], [128], [138].

3. In the statement of the alternative discrepancy principle we can put $\mu_\delta^\kappa(u_\delta, A) = 0$; moreover, Theorem 1.7 remains valid [64]. However, in this case, the choice of

regularization parameter in accordance with the alternative discrepancy principle is guaranteed only for those $u_\delta \in U$ for which there is a $\overline{x} \in D$ such that $\|AV\overline{x}-u_\delta\| \le \delta$.

About choosing the regularization parameter in nonlinear problems, see [7], [46], [47], [154], [178]. The further development of constructive means for choosing the regularization parameter in nonlinear problems took place in the papers [112], [114]–[117] of A.S. Leonov, and he has considered the case when the operator is approximately given.

9. Incompatible ill-posed problems

We return to the problem statement considered at the beginning of §1; however we will assume that the exact solution of (1.1), $Az = u$ with $z \in D \subseteq Z$, $u \in U$, Z, U Hilbert spaces, for $u = \overline{u}$ is, possibly, nonexistent but that there is an element $\overline{z} \in Z$ at which the discrepancy takes its minimum:

$$\overline{\mu} = \inf_{z \in D} \|Az - \overline{u}\| \ge 0$$

($\overline{\mu}$ is the incompatibility measure of equation (1.1) with exactly specified data $\overline{u}$, A) If $\overline{\mu} > 0$, then such an element $\overline{z} = \mathrm{argmin}_{z \in D} \|Az - \overline{u}\|$ ($\|A\overline{z} - \overline{u}\| = \overline{\mu}$, $\overline{z} \in D$) is called a *pseudosolution*. If the set of pseudosolutions $\overline{Z} = \{z\colon z \in D, \|Az - \overline{u}\| = \overline{\mu}\}$ consists of more than one element, then it is easily seen to be convex and closed, and hence there is a unique element $\overline{z} \in \overline{Z}$ nearest to zero. Such an element is called the *normal pseudosolution*:

$$\overline{z} = \mathrm{argmin}_{z \in \overline{Z}} \|z\|. \tag{1.43}$$

If the pseudosolution $\overline{z}$ is unique, we will also call it the normal pseudosolution; i $\overline{\mu} = 0$ (equation (1.1) is compatible), then a pseudosolution is also a solution and the normal pseudosolution is also the normal solution.

We will consider the, rather common, problem of constructing an approximation to the normal pseudosolution of the incompatible equation (1.1) with approximate data $\{A_h, u_\delta, \eta\}$; $\eta = (\delta, h)$, $\delta > 0$, $h \ge 0$, $\|u_\delta - \overline{u}\| \le \delta$, $\|A_h - A\| \le h$, a particular case o which is $\overline{\mu} = 0$.

We can easily show that the algorithm constructed above (§2) is not regularizing if $\overline{\mu} > 0$. Consider the following example [100]: $D = Z = U = \mathbf{R}^2$, $z = \begin{pmatrix} x \\ y \end{pmatrix}$

Equation (1.1) is taken to be a system of linear algebraic equations. The exact system has the form

$$\begin{array}{l} x = 0 \\ x = 1 \end{array}, \text{ i.e. } A = \begin{pmatrix} 1 & 0 \\ 1 & 0 \end{pmatrix}, \quad \overline{u} = \begin{pmatrix} 0 \\ 1 \end{pmatrix}.$$

The approximate system has the form

$$\begin{array}{l} x + \sigma y = \delta \\ x + 2\sigma y = 1 \end{array}, \text{ i.e. } A = \begin{pmatrix} 1 & \sigma \\ 1 & 2\sigma \end{pmatrix}, \quad \overline{u} = \begin{pmatrix} 0 \\ 1 \end{pmatrix}, \text{ where } |\sigma| \le \frac{h}{\sqrt{5}},$$

$$\|A - A_h\| \le h, \quad u_\delta = \begin{pmatrix} \delta \\ 1 \end{pmatrix}, \quad \|u_\delta - \overline{u}\| = \delta, \quad \delta > 0.$$

The incompatibility measure of the exact system is $\bar{\mu} = 1/\sqrt{2}$, the normal pseudosolution is $\bar{z} = \begin{pmatrix} 1/2 \\ 0 \end{pmatrix}$. The incompatibility measure of the perturbed system is $\mu_\eta = \inf_z \|A_h z - u_\delta\| = 0$ for all $\sigma \neq 0$, $\delta \geq 0$, since the matrix (of) A_h is nondegenerate. Consider the generalized discrepancy method, which consists in solving the problem (§3):

find

$$\inf \|z\|, \quad z \in Z_\eta = \left\{ z : z \in D, \, \|A_h z - u_\delta\|^2 \leq \left(\delta + h\|z\|\right)^2 + \mu_\eta^2(u_\delta, A_h) \right\},$$

or, in this case, $z = \begin{pmatrix} x \\ y \end{pmatrix}$,

$$Z_\eta = \left\{ z : (x + \sigma y - \delta)^2 + (x + 2\sigma y - 1)^2 \leq \left[\delta + h(x^2 + y^2)^{1/2}\right]^2 \right\}.$$

For sufficiently small h, δ we have $\bar{z} \notin Z_\eta$, since, when assuming the contrary, we arrive at the false inequality $1/2 \leq \delta(1 - \delta) + (\delta + h/2)^2$ whose righthand side can be made arbitrary small. It can be readily seen that it is impossible to exhibit elements $\tilde{z}_\eta \in Z_\eta$ such that $\tilde{z}_\eta \to \bar{z}$ as $\eta \to 0$. Thus, the solution obtained by the generalized discrepancy method (as well as by the generalized discrepancy principle, which is equivalent to it) does not converge to $\bar{z}$ as $\delta, h \to 0$. However, the algorithm constructed in §2 can be readily generalized to the case of incompatible operator equations, if instead of the incompatibility measure $\mu_\eta(u_\delta, A_h) = \inf_{z \in D} \|A_h z - u_\delta\|$ we use the following lower bound as incompatibility measure [35], [111]:

$$\hat{\mu}_\eta(u_\delta, A_h) = \inf_{z \in D} \left(\delta + h\|z\| + \|A_h z - u_\delta\|\right). \tag{1.44}$$

LEMMA 1.14 ([35]). *Under the above stated conditions, $\hat{\mu}_\eta(u_\delta, A_h) \geq \bar{\mu}$; moreover, $\lim_{\eta \to 0} \hat{\mu}_\eta(u_\delta, A_h) = \bar{\mu}$.*

PROOF. Since for arbitrary $z \in D$,

$$\bar{\mu} = \inf_{z \in D} \|A_h z - u_\delta\| \leq \|(A_h - A)z\| + \|\bar{u} - u_\delta\| + \|A_h z - u_\delta\| \leq$$

$$\leq h\|z\| + \delta + \|A_h z - u_\delta\|,$$

we have

$$\bar{\mu} \leq \inf_{z \in D} \left(\delta + h\|z\| + \|A_h z - u_\delta\|\right) = \hat{\mu}_\eta(u_\delta, A_h).$$

Further, let $\bar{z} \in D$ be such that $\|A\bar{z} - \bar{u}\| = \bar{\mu}$. Then

$$\bar{\mu} \leq \hat{\mu}_\eta(u_\delta, A_h) = \inf_{z \in D} \left(\delta + h\|z\| + \|A_h z - u_\delta\|\right) \leq$$

$$\leq \delta + h\|\bar{z}\| + \|A_h \bar{z} - u_\delta\| \leq$$

$$\leq \delta + h\|\bar{z}\| + \|(A - A_h)\bar{z}\| + \|u_\delta - \bar{u}\| + \|A\bar{z} - \bar{u}\| \leq$$

$$\leq 2\left(\delta + h\|\bar{z}\|\right) + \bar{\mu},$$

i.e.

$$\bar{\mu} \leq \hat{\mu}_\eta(u_\delta, A_h) \leq \bar{\mu} + 2\left(\delta + h\|\bar{z}\|\right);$$

this implies the second assertion of the lemma. $\square$

We denote, as before, the extremal of the functional $M^\alpha[z]$ (see (1.2)) by z_η^α and introduce the generalized discrepancy [101]

$$\hat{\rho}_\eta(\alpha) = \|A_h z_\eta^\alpha - u_\delta\|^2 - \left(\delta + h\|z_\eta^\alpha\| + \hat{\mu}_\eta(u_\delta, A_h)\right)^2. \tag{1.45}$$

It differs a bit from the generalized discrepancy introduced in §2; in particular, μ_η has been replaced by $\hat{\mu}_\eta$. If $\hat{\mu}_\eta$ is computed with error $\kappa \geq 0$, then, similarly as in §2, we can introduce the generalized discrepancy

$$\hat{\rho}_\eta^\kappa(\alpha) = \|A_h z_\eta^\alpha - u_\delta\|^2 - \left(\delta + h\|z_\eta^\alpha\| + \hat{\mu}^\kappa\eta(u_\delta, A_h)\right)^2,$$

where

$$\hat{\mu}_\eta(u_\delta, A_h) \leq \hat{\mu}_\eta^\kappa(u_\delta, A_h) \leq \hat{\mu}_\eta(u_\delta, A_h) + \kappa.$$

Note that in accordance with Lemma 1.14, $\hat{\mu}_\eta^\kappa(u_\delta, A_h) \geq \overline{\mu}$, since $\kappa \geq 0$ and $\hat{\mu}_\eta^\kappa(u_\delta, A_h) \to \overline{\mu}$ as $\eta \to 0$, if κ is matched with η in the sense that $\kappa = \kappa(\eta) \to 0$ as $\eta \to 0$. Similar to the derivation of the corollary to Lemma 1.3 we can obtain the following properties of the generalized discrepancy $\hat{\rho}_\eta^\kappa(\alpha)$:

1) $\hat{\rho}_\eta^\kappa(\alpha)$ is continuous and monotonically nondecreasing for $\alpha > 0$;

2) $\lim_{\alpha \to \infty} \hat{\rho}_\eta^\kappa(\alpha) = \|u_\delta\|^2 - \left(\delta + \hat{\mu}_\eta^\kappa(u_\delta, A_h)\right)^2$;

3) $\lim_{\alpha \downarrow 0} \hat{\rho}_\eta^\kappa(\alpha) \leq \left(\mu_\eta^\kappa(u_\delta, A_h)\right)^2 - \left(\delta + \hat{\mu}_\eta^\kappa(u_\delta, A_h)\right)^2 < 0$, since $\mu_\eta(u_\delta, A_h) < \mu_\eta^\kappa(u_\delta, A_h)$;

4) suppose the condition

$$\|u_\delta\| > \delta + \hat{\mu}_\eta^\kappa(u_\delta, A_h) \tag{1.46}$$

holds, then there is an $\alpha^* > 0$ such that

$$\hat{\rho}_\eta^\kappa(\alpha^*) = 0. \tag{1.47}$$

The last equality is equivalent to

$$\|A_h z_\eta^{\alpha^*} - u_\delta\| = \delta + h\|z_\eta^{\alpha^*}\| + \hat{\mu}_\eta^\kappa(u_\delta, A_h), \tag{1.48}$$

and, moreover, the element $z_\eta^{\alpha^*}$ is nonzero and uniquely defined.

We state the following modification of the generalized discrepancy principle: For data $\{A_h, u_\delta, \eta\}$, compute

$$\hat{\mu}_\eta^\kappa(u_\delta, A_h) = \inf_{z \in D} \left(\delta + h\|z\| + \|A_h z - u_\delta\|\right) + \kappa = \hat{\mu}_\eta(u_\delta, A_h) + \kappa,$$

$$\kappa \geq 0, \qquad \kappa = \kappa(\eta) \to 0 \quad \text{as } \eta \to 0$$

(see below for more about the computation of $\hat{\mu}_\eta^\kappa(u_\delta, A_h)$).

If the condition (1.46) is not satisfied, then we can take $z_\eta = 0$ as approximation to the normal pseudosolution of (1.1). If (1.46) is satisfied, then we can find an α^* satisfying (1.47). We then take $z_\eta = z_\eta^{\alpha^*}$ as approximation of the normal pseudosolution of (1.1).

THEOREM 1.8. $\lim_{\eta \to 0} z_\eta = \overline{z}$, where $\overline{z}$ is the normal pseudosolution of (1.1), i.e. the algorithm constructed above is a regularizing algorithm for finding the normal pseudosolution of (1.1).

PROOF. If $\overline{z} = 0$, then

$$\|u_\delta\| = \|A\overline{z} - \overline{u} + \overline{u} - u_\delta\| \leq \overline{\mu} + \delta \leq \hat{\mu}_\eta^\kappa(u_\delta, A_h) + \delta,$$

i.e. condition (1.46) is not satisfied and, in accordance with the generalized discrepancy principle, $z_\eta = 0$ and hence $z_\eta \to \overline{z}$ as $\eta \to 0$.

Suppose now that $\overline{z} \neq 0$. Then

$$\|\overline{u}\| > \overline{\mu}$$

(the assumption $\|\overline{u}\| < \overline{\mu}$ contradicts the definition of incompatibility measure, since $\|Az' - \overline{u}\| = \|\overline{u}\| < \overline{\mu}$ for $z' = 0 \in D$, while the assumption $\|\overline{u}\| = \overline{\mu}$ contradicts the fact that $\overline{z} \neq 0$ as well as the uniqueness of the normal pseudosolution of (1.1)).

Since $\delta + \hat{\mu}_\eta^\kappa(u_\delta, A_h) \to \overline{\mu}$ as $\eta \to 0$, (1.48) implies that condition (1.46) holds, at least for sufficiently small η. As in the proof of Theorem 1.2, we assume that $z_\eta^{\alpha^*(\eta)} \not\to \overline{z}$. This means that there are an $\epsilon > 0$ and a sequence $\eta_k \to 0$ such that $\|\overline{z} - z_{\eta_k}^{\alpha_k^*}\| \geq \epsilon$, $\alpha_k^* = \alpha^*(\eta_k)$ being the solution of (1.47).

By the extremal properties of $z_{\eta_k}^{\alpha_k^*} \in D$ $(z_{\eta_k} = z_{\eta_k}^{\alpha^*})$,

$$\|A_{h_k} z_{\eta_k} - u_{\delta_k}\|^2 + \alpha_k^* \|z_{\eta_k}\|^2 \leq \|A_{h_k} \overline{z} - u_{\delta_k}\|^2 + \alpha_k^* \|\overline{z}\|^2,$$

whence, using (1.48) and the results of Lemma 1.14,

$$\left(\delta_k + h_k \|z_{\eta_k}\| + \overline{\mu}\right)^2 + \alpha_k^* \|z_{\eta_k}\|^2 \leq \left(\delta_k + h_k \|z_{\eta_k}\| + \hat{\mu}_\eta^\kappa(u_{\delta_k}, A_{h_k})\right)^2 + \alpha_k^* \|z_{\eta_k}\|^2 =$$
$$= \|A_{h_k} z_{\eta_k} - u_{\delta_k}\|^2 + \alpha_k^* \|z_{\eta_k}\|^2 \leq$$
$$\leq \left(\delta_k + h_k \|\overline{z}\| + \overline{\mu}\right)^2 + \alpha_k^* \|\overline{z}\|^2. \tag{1.49}$$

(Here we have used the inequality

$$\|A_h \overline{z} - u_\delta\| = \|(A_h - A)\overline{z} + \overline{u} - u_\delta + A\overline{z} - \overline{u}\| \leq \delta + h\|\overline{z}\| + \overline{\mu}.)$$

Relation (1.49) and the strong monotonicity of the real-valued function $f(x) = (A + Bx)^2 + Cx^2$ of the real variable $x > 0$, $A = \delta_k + \overline{\mu} \geq 0$, $B = h_k \geq 0$, $C = \alpha_k^* > 0$, imply that

$$\|z_{\eta_k}\| \leq \|\overline{z}\|. \tag{1.50}$$

Taking into account, as in the proofs of Theorem 1.1 and Theorem 1.2, that $z_{\eta_k} \xrightarrow{\text{wk}} z^*$, $z^* \in D$, we obtain the inequality

$$\|z^*\| \leq \liminf_{k \to \infty} \|z_{\eta_k}\| \leq \limsup_{k \to \infty} \|z_{\eta_k}\| \leq \|z^*\|. \tag{1.51}$$

Now, to finish the proof and to obtain a contradiction with the assumption made at the beginning of the theorem, it remains to show that $z^* = \overline{z}$.

Since

$$\|Az_{\eta_k} - \overline{u}\| \leq \delta_k + h_k \|z_{\eta_k}\| + \|A_{h_k} z_{\eta_k} - u_{\delta_k}\| \leq$$
$$\leq 2 \left(\delta_k + h_k \|z_{\eta_k}\| \right) + \hat{\mu}_\eta(u_{\delta_k}, A_{h_k}) \leq$$
$$\leq 2 \left(\delta_k + h_k \|\overline{z}\| \right) + \hat{\mu}_\eta(u_{\delta_k}, A_{h_k}) \longrightarrow \overline{\mu}$$

as $\eta \to 0$, by the weak lower semicontinuity of $\|Az - \overline{u}\|$ with respect to z and the definition of incompatibility measure, we have $\|Az^* - \overline{u}\| = \overline{\mu}$. By (1.51) and the uniqueness of the normal pseudosolution, $z^* = \overline{z}$. Further, as in the proof of Theorem 1.1, (1.51) implies that $\|z_{\eta_k}\| \to \|\overline{z}\|$. Taking into account that $z_{\eta_k} \xrightarrow{\text{wk}} \overline{z}$ and the fact that the Hilbert space Z has the H-property, we obtain $z_{\eta_k} \to \overline{z}$. $\square$

REMARKS. 1. We stress that the modification of the generalized discrepancy principle given above is a regularizing algorithm independently of the fact whether or not equation (1.1) is compatible.

2. We can prove, in a way similar to Theorem 1.5, that the modification of the generalized discrepancy principle given above is equivalent to the generalized discrepancy method, which consists, in the case under consideration, of solving the following extremal problem with constraints:

find

$$\inf \|z\|, \qquad z \in \left\{ z \in D : \|A_h z - u_\delta\| \leq \delta + \hat{\mu}_\eta^\kappa(u_\delta, A_h) + h\|z\| \right\}.$$

For other approaches to the solution of linear ill-posed incompatible problems, see [45], [135], [138].

The generalized discrepancy method can be used also for the solution of nonlinear incompatible ill-posed problems, provided a bound on the incompatibility measure can be constructed in a corresponding manner [102]. Suppose that the assumptions stated at the beginning of §8 hold. Consider equation (1.39), which is, in general, incompatible. Let the incompatibility measure

$$\overline{\mu} = \inf_{z \in D} \|AVx - \overline{u}\| \tag{1.52}$$

correspond to the exact data $\{A, \overline{u}\}$, with set of pseudosolutions

$$\overline{X} = \left\{ \overline{x} : \ \overline{x} \in D, \ \|AV\overline{x} - \overline{u}\| = \overline{\mu} \right\} \neq \emptyset.$$

Similar to the proof of Remark 2 to the above theorem we can easily prove that $\overline{X}$ contains a nonempty set of normal pseudosolutions

$$\overline{X}_N = \left\{ \overline{x}_N : \ \overline{x}_N = \underset{x \in \overline{X}}{\operatorname{argmin}} \|x\| \right\} = \underset{x \in \overline{X}}{\operatorname{Argmin}} \|x\|.$$

We now construct an upper bound for the incompatibility measure from approximate data $\{h, u_\delta, \eta\}$:

$$\hat{\mu}_\eta(u_\delta, A_h) = \inf_{z \in D} \left\{ \delta + \psi(h, \|Vx\|) + \|A_h Vx - u_\delta\| \right\} \tag{1.53}$$

(the requirements for $\psi(h, y)$ were stated at the beginning of §8).

LEMMA 1.15. *Under the conditions stated in* §8 *(regarding the spaces* X, Z, U, *the operators* A, A_h *and the function* $\psi(h, y)$*) we have* $\hat{\mu}_\eta(u_\delta, A_h) \geq \overline{\mu}$. *Moreover,* $\lim_{\eta \to 0} \hat{\mu}(u_\delta, A_h) = \overline{\mu}$.

PROOF. Similar to the proof of Lemma 1.14 we note that for each $x \in D$,

$$\overline{\mu} = \inf_{x \in D} \|AVx - \overline{u}\| \leq \|AVx - \overline{u}\| =$$
$$= \|AVx - A_hVx + u_\delta - \overline{u} + A_hVx - u_\delta\| \leq$$
$$\leq \psi(h, \|Vx\|) + \delta + \|A_hVx - u_\delta\|.$$

Hence

$$\overline{\mu} \leq \inf_{x \in D} \left(\delta + \psi(h, \|Vx\|) + \|A_hVx - u_\delta\|\right) = \hat{\mu}_\eta(u_\delta, A_h).$$

Further, let $\overline{x} \in \overline{X} \subseteq D$, i.e. $\|AV\overline{x} - \overline{u}\| = \overline{\mu}$. Then

$$\overline{\mu} \leq \hat{\mu}_\eta(u_\delta, A_h) = \inf_{x \in D} \left(\delta + \psi(h, \|Vx\|) + \|A_hVx - u_\delta\|\right) \leq$$
$$\leq \delta + \psi(h, \|V\overline{x}\|) + \|A_hV\overline{x} - u_\delta\| \leq$$
$$\leq \delta + \psi(h, \|V\overline{x}\|) + \|A_hV\overline{x} - AV\overline{x}\| + \|\overline{u} - u_\delta\| + \|AV\overline{x} - \overline{u}\| \leq$$
$$\leq 2\left(\delta + \psi(h, \|V\overline{x}\|)\right) + \overline{\mu} \longrightarrow \overline{\mu}$$

as $\eta \to 0$, which proves the second assertion of the lemma. $\square$

If we can compute a bound for the incompatibility measure with error $\kappa \geq 0$, then it is natural to consider $\hat{\mu}_\eta^\kappa(u_\delta, A_h)$:

$$\hat{\mu}_\eta(u_\delta, A_h) \leq \hat{\mu}_\eta^\kappa(u_\delta, A_h) \leq \hat{\mu}_\eta(u_\delta, A_h) + \kappa,$$

where we will assume that $\kappa = \kappa(\eta) \to 0$ as $\eta \to 0$. In this case Lemma 1.15 is true also for $\hat{\mu}_\eta^\kappa(u_\delta, A_h)$. We will now consider the extremal problem
find

$$\inf_{x \in X} \|x\|, \qquad X_\eta = \left\{x : x \in D, \|A_hVx - u_\delta\| \leq \delta + \psi(h, \|Vx\|) + \hat{\mu}_\eta^\kappa(u_\delta, A_h)\right\}. \tag{1.54}$$

It is obvious that $X_\eta \neq \emptyset$, since it contains at least $\overline{X}$, $\overline{X} \subseteq X_\eta$. Indeed, for any element $\overline{x} \in \overline{X} \subseteq D$,

$$\|A_hV\overline{x} - u_\delta\| = \|A_hV\overline{x} - AV\overline{x} + \overline{u} - u_\delta + AV\overline{x} - \overline{u}\| \leq$$
$$\leq \psi(h, \|V\overline{x}\|) + \delta + \overline{\mu} \leq$$
$$\leq \delta + \psi(h, \|V\overline{x}\|) + \hat{\mu}_\eta^\kappa(u_\delta, A_h).$$

We now take an arbitrary element $\overline{x}_N$. Clearly, $\overline{x}_N \in \overline{X}_N \subseteq \overline{X}$, therefore the problem (1.54) is equivalent to the problem:
find

$$\inf \|x\|, \qquad x \in X_\eta \cap \overline{S}(0, \|\overline{x}_N\|), \tag{1.55}$$

where $\overline{S}(0, \|\overline{x}_N\|)$ is the closed unit ball in X with center at zero and radius $\|\overline{x}_N\|$.

Literally repeating the proof of Lemma 1.9 (using only the fact that the definition of X_η involves a bound $\hat{\mu}_\eta^\kappa(u_\delta, A_h)$ for the incompatibility measure), we can easily prove that the set $X_\eta \cap \overline{S}(0, \|\overline{x}_N\|)$ is weakly compact in X, and hence the problem (1.55) is solvable. Let X_η^* be its set of solutions.

THEOREM 1.9. X_η^* β-converges to $\overline{X}_N$ as $\eta \to 0$.

PROOF. β-convergence of X_η^* to $\overline{X}_N$ as $\eta \to 0$ means that the β-deviation

$$\beta(X_\eta^*, \overline{X}_N) = \sup_{x_\eta^* \in X_\eta^*} \inf_{\overline{x}_N \in \overline{X}_N} \|x_\eta^* - \overline{x}_N\| \longrightarrow^{\eta \to 0} 0.$$

The proof of the theorem goes by contradiction. Suppose we can find a sequence $\eta_n \to 0$, elements $x_{\eta_n} \in X_{\eta_n}^*$ and a number $\epsilon > 0$ such that $\|x_{\eta_n} - \overline{x}_N\| \geq \epsilon$ for all $\overline{x}_N \in \overline{X}_N$. Put $x_{\eta_n} \equiv x_n$. Since (1.55) implies $\|x_n\| \leq \|\overline{x}_N\|$ and X is a Hilbert space, we can extract a subsequence x_{n_k} from x_n that converges weakly to $x^* \in D \cap \overline{S}(0, \|\overline{x}_N\|)$. Moreover,

$$\|AVx_{n_k} - \overline{u}\| \leq \|A_{h_{n_k}} Vx_{n_k} - u_{\delta_{n_k}}\| + \|AVx_{n_k} - A_{h_{n_k}} Vx_{n_k}\| + \|u_{\delta_{n_k}} - \overline{u}\| \leq$$
$$\leq 2\left(\delta_{n_k} + \psi(h_{n_k}, \|Vx_{n_k}\|)\right) + \mu_{\eta n_k}^{\kappa_{n_k}}(u_{\delta_{n_k}}, A_{h_{n_k}}), \quad \kappa_{n_k} \equiv \kappa(\eta_{n_k}).$$

The sequence x_{n_k} converges weakly, therefore the sequence $\|Vx_{n_k}\|$ is bounded: $0 \leq \|Vx_{n_k}\| \leq C$. Transition to the limit as $k \to \infty$ and using the properties of $\psi(h, y)$, the compactness of AV and Lemma 1.15, we find $\|AVx^* - \overline{u}\| = \overline{\mu}$. Since $x^* \in D$, also $x^* \in \overline{X}$. Further, $\|x\|$ is weakly lower semicontinuous, so

$$\|x^*\| \leq \liminf_{k \to \infty} \|x_{n_k}\| \leq \limsup_{k \to \infty} \|x_{n_k}\| \leq \|\overline{x}_N\|,$$

i.e. $x^* \in \overline{X}_N$ and $\lim_{k \to \infty} \|x_{n_k}\| = \|x^*\|$. Since X is a Hilbert space, $\lim_{k \to \infty} x_{n_k} = x^* \in \overline{X}_N$, which leads to a contradiction. □

COROLLARY. *Suppose the normal pseudosolution $\overline{x}_N$ is unique. Then $x_\eta \to \overline{x}_N$ as $\eta \to 0$, where x_η is an arbitrary element in X_η^*.*

For the use of the generalized discrepancy principle in the solution of nonlinear incompatible ill-posed problems with approximately given operator, see [116], [117].

We conclude this section with the problem of computing a lower bound $\hat{\mu}_\eta(u_\delta, A_h)$ for the incompatibility measure when solving linear ill-posed problems. We introduce the functional

$$\Phi_\eta(z) = \delta + h\|z\| + \|A_h z - u_\delta\|. \tag{1.56}$$

Then

$$\hat{\mu}_\eta(u_\delta, A_h) = \inf_{z \in D} \Phi_\eta(z). \tag{1.57}$$

We note some elementary properties of $\Phi_\eta(z)$:

1) $\Phi_\eta(z)$ is a convex continuous functional.
2) $\Phi_\eta(z)$ is everywhere Fréchet differentiable, except at $z = 0$ and $z = A_h^{-1} u_\delta$ (if the latter element exists).

3) if $h > 0$, then the problem (1.57) is solvable, i.e. there is a (possibly nonunique) element $\hat{z}_\eta \in D$ such that $\hat{\mu}_\eta(u_\delta, A_h) = \Phi_\eta(\hat{z})$.

We prove the last property. Consider a sequence $\{z_n\}_{n=1}^\infty$, $z_n \in D$, $\Phi_\eta(z_n) \to \inf_{z \in D} \Phi_\eta(z)$. Then $\Phi_\eta(z_n) \le \Phi_\eta(z_1)$, and problem (1.57) is equivalent to:
find
$$\hat{\mu}_\eta(u_\delta, A_h) = \inf \Phi_\eta(z), \qquad z \in D \cap \overline{S}(0, R),$$
where $\overline{S}(0, R)$ is the closed ball with center at zero and radius $R = (1/h)\Phi_\eta(z_1)$.

The set $D \cap \overline{S}(0, R)$ is convex, closed and bounded. Consequently, we can extract a subsequence $\{z_{n_k}\}$ from $\{z_n\} \subset D \cap \overline{S}(0, R)$ that converges weakly to an element $z^* \in D \cap \overline{S}(0, R)$. Since $\Phi_\eta(z_{n_k}) \to \inf_{z \in D} \Phi_\eta(z)$ and the functional $\Phi_\eta(z)$ is weakly lower semicontinuous, transition to the limit as $n_k \to \infty$ gives $\Phi_\eta(z^*) = \inf_{z \in D} \Phi_\eta(z)$, i.e. we can take z^* as $\hat{z}_\eta$.

So, the problem of finding $\hat{\mu}_\eta(u_\delta, A_h)$ is a convex programming problem, for the solution of which we have efficient numerical methods. For $D = Z$ an algorithm for efficiently estimating the incompatibility measure can be found in [110], [111]. We describe this algorithm here.

LEMMA 1.16 ([110]). *Let $h > 0$. The functional $\Phi_\eta(z)$ attains its minimum at $\hat{z}_\eta = 0$ if and only if the following inequality holds:*

$$h\|u_\delta\| \ge \|A_h^* u_\delta\|. \tag{1.58}$$

PROOF. We write $\Phi_\eta(z)$ as a sum $\Phi_\eta(z) = \Phi_1(z) + \Phi_2(z)$, where $\Phi_1(z) = \|A_h z - u_\delta\|$, $\Phi_2(z) = h\|z\| + \delta$. If $u_\delta = 0$, then it is clear that $\hat{z}_\eta = 0$ is the unique minimum point of $\Phi_\eta(z)$, and (1.58) is satisfied. For $u_\delta \ne 0$ we see, using the fact that $\Phi_1(z), \Phi_2(z)$ are continuous convex functionals on Z, $\Phi_1(z)$ is Fréchet differentiable in a neighborhood of $z = 0$ and $\Phi_1'(0) = -A_h^* u_\delta/\|u_\delta\|$, that in accordance with [110], $\hat{z}_\eta = 0$ is a minimum point of $\Phi_\eta(z)$ if and only if

$$\left(\Phi_1'(0), z - 0\right) + \Phi_2(z) - \Phi_2(0) \ge 0 \qquad \forall z \in Z,$$

or:

$$\left(-A_h^* \frac{u_\delta}{\|u_\delta\|}, z\right) + h\|z\| + \delta - \delta \ge 0 \qquad \forall z \in Z.$$

Hence,

$$h\|z\| \ge \left(A_h^* \frac{u_\delta}{\|u_\delta\|}, z\right) \qquad \forall z \in Z,$$

or:

$$h \ge \left(A_h^* \frac{u_\delta}{\|u_\delta\|}, \frac{z}{\|z\|}\right) \qquad \forall z \in Z. \tag{1.59}$$

The maximum of the righthand side of (1.59) is clearly attained at $z = A_h^* u_\delta$, which proves the lemma. $\square$

REMARK. In a numerical realization, the condition (1.58) can be readily verified.

LEMMA 1.17 ([110]). *If the functional $\Phi_\eta(z)$ for $h > 0$ attains its minimum in at least two distinct elements $z_1, z_2 \in Z$, then the following assertions are valid:*

1) the equation $A_h z = u_\delta$ is solvable;
2) $\Phi_\eta(z)$ attains its minimum on the interval $[0, \hat{z}]$, where $\hat{z}$ is the normal solution of the equation $A_h z = u_\delta$;
3) $\min_{z \in Z} \Phi_\eta(z) = \|u_\delta\| + \delta + h\|\hat{z}\| + \delta$.

PROOF. Since $\Phi_\eta(z)$ is a convex continuous functional, its set of extremals $Z_{\min}$ is convex, closed (see, e.g., [71]) and, as has been proved above, bounded. By assumption, there are $z_1, z_2 \in Z_{\min}$, $z_1 \neq z_2$. For the sake of being specific we assume that $z_2 \neq 0$. Since $Z_{\min}$ is convex, the interval joining z_1 and z_2, i.e. $[z_1, z_2] = \{z: z = \lambda z_1 + (1 - \lambda)z_2, \lambda \in [0,1]\}$ also belongs to $Z_{\min}$. Therefore, for any $\lambda \in [0, 1]$,

$$\Phi_\eta(\lambda z_1 + (1 - \lambda)z_2) = \min_{z \in Z} \Phi_\eta(z) =$$
$$= \delta + h\|\lambda z_1 + (1 - \lambda)z_2\| + \left\|A_h\left(\lambda z_1 + (1 - \lambda)z_2\right) - u_\delta\right\| =$$
$$= \delta + \lambda h\|z_1\| + (1 - \lambda)h\|z_2\| +$$
$$+ \lambda\|A_h z_1 - u_\delta\| + (1 - \lambda)\|A_h z_2 - u_\delta\|. \tag{1.60}$$

This equality and the convexity of $\|z\|$ and $\|A_h z - u_\delta\|$ imply that for any $\lambda \in [0, 1]$,

$$\|\lambda z_1 + (1 - \lambda)z_2\| = \lambda\|z_1\| + (1 - \lambda)\|z_2\|, \tag{1.61}$$
$$\|\lambda(A_h z_1 - u_\delta) + (1 - \lambda)(A_h z_2 - u_\delta)\| = \lambda\|A_h z_1 - u_\delta\| + (1 - \lambda)\|A_h z_2 - u_\delta\|.$$

The proof of these identities can be readily obtained by contradiction. If we assume that for some $\lambda^* \in (0, 1)$ some equality in (1.61) does not hold, then (1.60) will not hold too.

By the strong convexity of the Hilbert spaces Z and U, there are numbers $\kappa \neq 1$ and β such that either

$$z_1 = \kappa z_2, \qquad A_h z_1 - u_\delta = \beta(A_h z_2 - u_\delta), \tag{1.62}$$

or

$$z_1 = \kappa z_2, \qquad A_h z_1 - u_\delta = 0 \quad (\text{i.e. } \beta = 0).$$

In the second case assertion 1) of the lemma holds. Consider the first case. We can easily prove by contradiction that $\beta \neq 1$. Indeed, let $\beta = 1$. So,

$$\Phi_\eta(z_1) = \Phi_\eta(z_2),$$

i.e.

$$\delta + h\|z_1\| + \|A_h z_2 - u_\delta\| = \delta + h\|z_2\| + \|A_h z_2 - u_\delta\|.$$

Using that $\beta = 1$ we thus obtain $\|z_1\| = |\kappa| \, \|z_2\|$, i.e. $|\kappa| = 1$. If $\kappa = 1$ we arrive at a contradiction with $z_1 \neq z_2$. If $\kappa = -1$ then $z_1 = -z_2$ and (1.62) implies $A_h z_1 = A_h z_2 = 0$, i.e. $z_1, z_2 \in \operatorname{Ker} A_h$. But $z_1, z_2 \in Z_{\min}$ if and only if $z_1 = z_2 = 0$, since $\|A_h z_1 - u_\delta\| = \|A_h z_2 - u_\delta\| = \|u_\delta\|$ and $\|z_1\| = \|z_2\| \geq 0$. But by requirement $z_1 \neq z_2$. So, $\beta \neq 1$. Equation (1.62) implies

$$\hat{z} = \frac{(\kappa - \beta)z_2}{1 - \beta}, \qquad A_h \hat{z} = u_\delta, \qquad \hat{z} \neq 0.$$

Assertion 1) of the lemma has been proved completely.

Obviously, $u_\delta \neq 0$ (since if $u_\delta = 0$, then $\Phi_\eta(z)$ attains its minimum at $z_1 = z_2 = 0$ only). Therefore $\kappa \neq \beta$. Put $\tau = (1 - \beta)/(\kappa - \beta)$, $t = \kappa\tau$. Then $z_2 = \tau\hat{z}$, $z_1 = \tau\hat{z}$ (if $A_h z_2 = u_\delta$, then we choose $\hat{z} = z_2$, $\tau = 1$, $\tau = \kappa$). Since $z_1 \neq z - 2$, we have $t \neq \tau$.

Note that

$$\Phi(z_1) = \delta + h\|z_1\| + \|A_h z_1 - u_\delta\| =$$
$$= \delta + |t|\,\|\hat{z}\| + \|t A_h \hat{z} - t u_\delta + t u_\delta - u_\delta\| =$$
$$= \delta + |t|\,\|\hat{z}\| + |t - 1|\,\|u_\delta\|.$$

Similarly,

$$\Phi_\eta(z_2) = \delta + |\tau|\,\|\hat{z}\| + |\tau - 1|\,\|u_\delta\|,$$

and

$$\Phi_\eta(z_1) = \Phi_\eta(z_2).$$

Substituting the expressions for z_1, z_2 with $\lambda = 1/2$ into (1.61), we see that after division by $\|\hat{z}\| \neq 0$ and $\|u_\delta\| \neq 0$:

$$|t + \tau| = |t| + |\tau|,$$
$$|t - 1 + \tau - 1| = |t - 1| + |\tau - 1|,$$

i.e. t, τ (and similarly $t - 1$, $\tau - 1$) are either both nonnegative or both nonpositive.

Suppose first that $t \geq 0$, $\tau \geq 0$, $t - 1 \leq 0$, $\tau - 1 \leq 0$. Then the equation $\Phi_\eta(z_1) = \Phi_\eta(z_2)$ implies that

$$(1 - t)\|u_\delta\| + th\|\hat{z}\| + \delta = (1 - \tau)\|u_\delta\| + \tau h\|\hat{z}\| + \delta,$$

i.e. $(t - \tau)h\|\hat{z}\| = (t - \tau)\|u_\delta\|$. Since $t \neq \tau$, this implies assertion 3) of the lemma.

If $t \geq 0$, $\tau \geq 0$, $t - 1 \geq 0$, $\tau - 1 \geq 0$ then

$$(t - 1)\|u_\delta\| + th\|\hat{z}\| + \delta = (\tau - 1)\|u_\delta\| + \tau h\|\hat{z}\| + \delta,$$

or $(t - \tau)h\|\hat{z}\| = (\tau - t)\|u_\delta\|$, which is impossible since $t \neq \tau$, $\|\hat{z}\| > 0$, $\|u_\delta\| > 0$.

The cases $t \leq 0$, $\tau \leq 0$, $t - 1 \geq 0$, $\tau - 1 \geq 0$ (when assertion 3) is true) and $t \leq 0$, $\tau \leq 0$, $t - 1 \leq 0$, $\tau - 1 \leq 0$ (which is impossible) are treated similarly. So, $\Phi_\eta(\hat{z}) = h\|\hat{z}\| + \delta = \min_{z \in Z} \Phi_\eta(z)$.

To finish the proof of the lemma it remains to show that $\hat{z}$ is the normal solution of $A_h z = u_\delta$. Indeed, let z' be a solution of $A_h z = u_\delta$ with $\|u'\| < \|\hat{z}\|$. Then

$$\Phi_\eta(z') = \delta + h\|z'\| < \Phi_\eta(\hat{z}) = \delta + h\|\hat{z}\| = \min_{z \in Z} \Phi_\eta(z),$$

which is impossible.

The fact that the interval $[0, \hat{z}] \subseteq Z_{\min}$ follows from the equation stated in 3),

$$\Phi_\eta(0) = \delta + \|u_\delta\| = \Phi_\eta(\hat{z}) = \delta + h\|\hat{z}\|,$$

and the convexity of $Z_{\min}$. $\square$

COROLLARY. *If the inequality (1.58) is not satisfied, then the problem of minimizing* $\Phi_\eta(z)$ *for* $h > 0$ *has a unique solution* $\hat{z}_\eta \neq 0$. *In this case, if* $A_h \hat{z}_\eta \neq u_\delta$, *then* $\Phi_\eta(z)$ *is differentiable at* $\hat{z}_\eta$ *and* $\hat{z}_\eta$ *is a solution of the equation*

$$\Phi_\eta'(z) = \frac{A_h^* A_h z - A_h^* u_\delta}{\|A_h z - u_\delta\|} + h \frac{z}{\|z\|} = 0. \tag{1.63}$$

By the convexity of $\Phi_\eta(z)$, the condition (1.63) is a sufficient condition for a minimum of $\Phi_\eta(z)$.

LEMMA 1.18 ([110]). *If the condition (1.58) is not satisfied, then the problem of minimizing* $\Phi_\eta(z)$ *is equivalent to the problem of minimizing the functional* $M^\alpha[z]$ *given by (1.2) with regularization parameter chosen in accordance with the principle of least discrepancy estimate, i.e. we choose* $\hat{\alpha} \geq 0$ *such that* $\psi(\hat{\alpha}) = \min_{\alpha \geq 0} \psi(\alpha)$, *where the function* $\psi(\alpha)$ *is defined by* $\psi(\alpha) = \|A_h z_\eta^\alpha - u_\delta\| + h\|z_\eta^\alpha\|$.

PROOF. Lemma 1.17 and (1.58) imply that the solution $\hat{z}_\eta$ of the problem of minimizing $\Phi_\eta[z]$ exists and is unique. Let $\|\hat{z}_\eta\| = r > 0$. Clearly, $\hat{z}_\eta$ is the unique solution of the variational problem

find
$$\inf_{\|z\|^2 \leq r^2} \|A_h z - u_\delta\|^2.$$

In [90] it has been proved that there is an $\hat{\alpha} \geq 0$ such that $\hat{z}_\eta$ is an extremal of $M^{\hat{\alpha}}[z]$. Since $\hat{z}_\eta$ is the unique solution of the problem of minimizing $\Phi_\eta(z)$ on the whole space, it is of course also the solution of this problem on the set of extremals of $M^\alpha[z]$ for all $\alpha \geq 0$, i.e. $\psi(\hat{\alpha}) = \min_{\alpha \geq 0} \psi(\alpha)$. The continuity and strong monotonicity of $\|z_\eta^\alpha\|$ as a function of α (Lemma 1.5) imply that $\hat{\alpha}$ is unique. Note that $\hat{\alpha} = 0$ if $\lim_{\alpha \downarrow} \|z_\eta^\alpha\| = r = \|\hat{z}_\eta\|$ and, as has been shown in the proof of Theorem 1.3, in this case $\hat{z}_\eta$ is the normal pseudosolution of the equation $A_h z = u_\delta$. $\square$

To finish this section we prove the following theorem, in which we give a substantiation of the principle of least discrepancy estimate.

THEOREM 1.10 ([111]). *Let* $h > 0$. *The method of least discrepancy estimate is equivalent to the following generalized discrepancy principle. If (1.58) holds, we take* $\hat{z}_\eta = 0$. *In the opposite situation* $\hat{z}_\eta$ *can be found as the extremal of the functional* $M^\alpha[z]$ *with regularization parameter chosen from the condition that the function* $\psi(\alpha)$ *be minimal, for* $\alpha \geq 0$. *The function* $\psi(\alpha)$ *is continuously differentiable for* $\alpha > 0$, *has a unique local minimum* $\hat{\alpha}$ *which is also a global minimum on* $[0, \infty)$; $\hat{z}_\eta = z_\eta^{\hat{\alpha}}$. *If* $\hat{\alpha} \neq 0$, *then* $\hat{\alpha}$ *is the unique solution of the equation*

$$\alpha\|z_\eta^\alpha\| = h\|A_h z_\eta^\alpha - u_\delta\|. \tag{1.64}$$

PROOF. If (1.58) holds, then by Lemma 1.16, Lemma 1.17, $\hat{z}_\eta = 0$. Consider the second case. Recall that z_η^α is the solution of the Euler equation (1.5) for the functional $M^\alpha[z]$. Therefore

$$A_h^* A_h z_\eta^\alpha - A_h^* u_\delta = -\alpha z_\eta^\alpha. \tag{1.65}$$

In accordance with the results of §4 (Lemma 1.5), the function $\psi(\alpha) = \|A_h z_\eta^\alpha - u_\delta\| + h\|z_\eta^\alpha\|$ is continuously differentiable for $\alpha > 0$ and its derivative is:

$$\psi'(\alpha) = \frac{1}{2}\left((A_h^* A_h + \alpha E)^{-1} z_\eta^\alpha, z_\eta^\alpha\right)\left(\frac{\alpha}{\|A_h z_\eta^\alpha - u_\delta\|} - \frac{h}{\|z_\eta^\alpha\|}\right). \qquad (1.66)$$

If $\psi(\alpha)$ has a stationary point (a local extremum), then $\psi'(\alpha) = 0$ and (1.64) holds. Using (1.64), (1.65) we see that the sufficient condition (1.63) for $\Phi_\eta(z)$ to have an extremum holds, i.e. the extremal of $M^\alpha[z]$ corresponding to the regularization parameter α chosen as the solution of $\psi'(\alpha) = 0$ is an extremal of $\Phi_\eta(z)$. By Lemma 1.17, the extremal of $\Phi_\eta(z)$ is unique. By Lemma 1.18, there is a unique $\hat{\alpha} \geq 0$ such that $\hat{\alpha}$ is a global minimum of $\psi(\alpha)$ on $[0, \infty)$, and $\hat{z}_\eta = z_\eta^{\hat{\alpha}}$. Thus, $\psi(\alpha)$ does not have local minima. □

The results of the theorem have, basically, been laid down in a numerical algorithm that has been implemented on a computer.

CHAPTER 2

Numerical methods for the approximate solution of ill-posed problems on compact sets

In Chapter 1 we have discussed general questions concerning the construction of regularizing algorithms for solving a wide circle of ill-posed problems. In this Chapter we will consider in some detail the case when we a priori know that the exact solution of the problem belongs to a certain compact set. The idea of this approach was expressed already in 1943 [164].

In [85]–[87] the notion of quasisolution was introduced, it being an approximate solution of an ill-posed problem on a compact set. In these papers a whole series of theorems concerning existence, uniqueness and continuous dependence of the quasisolution on the righthand side were proved. We can give a sufficiently general approach to approximately finding quasisolutions [79], [87].

The papers [215], [216] are devoted to the construction of an approximate solution of ill-posed problems on compact sets. Here, the set of well-posedness is the image of the ball under the mapping by a compact operator. Questions related to this one have been considered in detail in [90], [108]. In these papers the set of well-posedness is some abstract set.

In our opinion, the following two questions are of interest:

PROBLEM 1. What 'qualitative' information regarding the exact solution will suffice to distinguish the set of well-posedness?

PROBLEM 2. How to effectively find the solution of an ill-posed problem, if this set has been distinguished?

For certain special problems (the inverse problem of potential theory, some problems in analytic function theory) a number of interesting results regarding the definition of the set of well-posedness have been obtained in [215], [85], [83], [84], [104]–[106], [147], [148].

In [74], [196] it was indicated for the first time that for a large class of inverse problems of mathematical physics, such as the inverse problem in geophysics, pattern recognition, diagnostic of plasmas, astrophysics, there is a priori information on the nature of the solution looked for (its monotonicity, convexity, etc.). Using such information we can construct efficient algorithms for solving ill-posed problems [49], [54], [65]–[67].

The application of methods for solving ill-posed problems based on the use of a priori information regarding the behavior of the unknown functions, has made it possible to obtain essential results for a large number of applied problems [6], [18]–[22], [69], [70], [72], [124], [194], [200].

1. Approximate solution of ill-posed problems on compact sets

Consider the ill-posed problem

$$Az = u, \qquad z \in Z, \quad u \in U. \tag{2.1}$$

Here, z, u belong to normed function spaces $Z[a, b]$, $U[c, d]$ and the operator $A \colon Z \to U$ is continuous. As usual, suppose that instead of the exact value $\overline{u} = A\overline{z}$ and exact A we are given an approximation u_δ and a continuous operator A_h such that $\|u_\delta - \overline{u}\| \leq \delta$, $\|A_h z - Az\| \leq \psi(h, \|z\|)$ (see Chapter 1, §8). If A and A_h are linear operators, $\psi(h, y) = hy$. Suppose our a priori information is that the exact solution $\overline{z}$ belongs to a compact set M and that A maps M onto $AM \subset U$. As before, let $\eta = (\delta, h)$.

Let $Z_M(\eta)$ be the set of elements $z_\eta \in M$ such that

$$\|A_h z_\eta - u_\delta\| \leq \psi(h, \|z_\eta\|) + \delta$$

(note that $Z_M(\eta)$ is defined for given values of η, A_h, u_δ). For each $\eta > 0$, $Z_m(\eta)$ is nonempty, since it contains $\overline{z}$.

LEMMA 2.1. *Let z_η be an arbitrary element in $Z_M(\eta)$. Then $z_\eta \to \overline{z}$ as $\eta \to 0$.*

PROOF. The set M is compact in Z, hence

$$\sup_{z \in M} \|z\| = C_0 < \infty.$$

Since $z_\eta \in Z_M(\eta)$, the following estimate holds:

$$\|Az_\eta - A\overline{z}\| \leq \|Az_\eta - A_h z_\eta\| + \|A_h z_\eta - u_\delta\| + \|u_\delta - \overline{u}\| \leq$$
$$\leq 2\big(\psi(h, \|z_\eta\|) + \delta\big).$$

The properties of $\psi(h, \|z\|)$, the continuity of A^{-1} on AM [95] and the last inequality imply the lemma. $\square$

Thus, if $\overline{z} \in M$, then to approximately solve the problem (2.1) we can take an arbitrary element $z_\eta \in Z_M(\eta)$. Moreover, if $h = 0$, then we can take an arbitrary element $z_\delta \in M$ such that $\|Az_\delta - u_\delta\| \leq \delta$ as approximate solution of (2.1).

We will consider the problem of the error in the approximate solution. It is natural to take

$$\epsilon(\eta) = \sup_z \|z_\eta^* - z\|, \qquad z \in \Big\{ z \colon z \in D, \ \|A_h z - u_\delta\| \leq \psi(h, \|z\|) + \delta \Big\},$$

as the error of a fixed approximation z_η^* of (2.1) with approximate righthand side u_δ and approximate operator A_h on the set of a priori constraints $D \subset Z$.

Clearly, $\|z_\eta^* - \overline{z}\| \leq \epsilon(\eta)$. Note that the problem of the error of the deviation of the approximate solution of the ill-posed problem (2.1) from the exact solution makes sense only if D is a compact set in Z, $D = M$. In this case $\epsilon(\eta) \to 0$ as $\eta \to 0$.

Thus, when it is a priori known that the exact solution $\bar{z}$ of the ill-posed problem (2.1) belongs to a compact set, we can not only indicate an approximate solution, but also find an error estimate of the approximation.

2. Some theorems regarding uniform approximation to the exact solution of ill-posed problems

In a large number of inverse problems of mathematical physics we have a priori information regarding the monotonicity of the solution looked for. This information turns out to be sufficient to construct stable algorithms for the approximate solution of equation (2.1).

Suppose we know a priori that the exact solution $\bar{z}(s)$ of the ill-posed problem (2.1) is a monotone function on an interval (for the sake of being specific we take it to be nonincreasing), and is bounded above and below by constants C_1 and C_2, respectively. Without loss of generality we may assume that $C_1 = C > 0$, $C_2 = 0$. Introduce the space $Z\downarrow_C$ of nonincreasing functions $z(s)$ that are bounded by the constants C and 0, i.e.

$$C \geq z(s) \geq 0, \qquad \forall s \in [a, b].$$

Let $z_1(s), z_2(s), \ldots$ be a sequence of functions in $Z\downarrow_C$. By Helly's theorem [95], there are a function $z(s) \in Z\downarrow_C$ and a sequence of indices n_k such that at each point $s \in [a, b]$,

$$\lim_{k \to \infty} z_{n_k}(s) = z(s).$$

Pointwise convergence and uniform boundedness imply $L_p[a, b]$-convergence, $p > 1$. Thus, $Z\downarrow_C$ is compact in L_p. We will assume that $Z = L_p[a, b]$, $p > 1$ and that the operator A is bijective from $Z\downarrow_C$ onto $AZ\downarrow_C$, i.e. $\|Az_1 - Az_2\| = 0$ for $z_1, z_2 \in Z\downarrow_C$ implies $z_1 \overset{Z}{=} z_2$.

Consider now the set $Z\downarrow_C (\eta)$ of elements $z \in Z\downarrow_C$ such that

$$\|A_h z - u_\delta\| \leq \psi(h, \|z\|) + \delta.$$

We will denote its elements by z_η.[1]

Lemma 2.1 implies that an arbitrary $z_\eta \in Z\downarrow_C (\eta)$ is an approximate solution of (2.1), and $z_\eta \to \bar{z}$ in L_p as $\eta \to 0$.

Suppose we know that the solution $\bar{z}(s) \in Z\downarrow_C$ is continuous on $[a, b]$; let $\eta_n \to 0$ as $n \to \infty$. Then the following stronger assertion holds:

THEOREM 2.1 ([74], [75]). *Let $\bar{z}(s) \in C[a, b]$, and let z_n be an arbitrary element in $Z\downarrow_C (\eta_n)$. Let $[\gamma, \rho]$ be an arbitrary fixed subinterval of (a, b). Then the sequence $z_n(s)$ converges uniformly on $[\gamma, \rho]$ to $\bar{z}(s)$ as $n \to \infty$.*

PROOF. We first prove that $z_n(s)$ converges to $\bar{z}(s)$ as $n \to \infty$ at each point $s \in (a, b)$. By Helly's theorem, we can always extract a subsequence $\{z_{n_k}\}$ from $\{z_n\}$ that converges at every $s \in [a, b]$, and thus also in Z, to a function $\bar{\bar{z}} \in Z\downarrow_C$. Let $A_{h'_k}, u_{\delta'_k}$ be subsequences of A_{h_n}, u_{δ_n}, respectively.

---

[1]The set $Z\downarrow_C (\eta)$ is defined for given elements A_h, u_δ, δ, h.

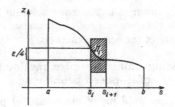

FIGURE 2.1

We can be easily convinced of the truth of the estimate

$$\|A\overline{z} - \overline{u}\| \le \|A\overline{z} - Az_{n_k}\| + \|Az_{n_k} - A_{h'_k} z_{n_k}\| + \|A_{h'_k} z_{n_k} - u_{\delta'_k}\| + \|u_{\delta'_k} - \overline{u}\| \le$$
$$\le \epsilon_k + 2 \left(\psi(h'_k, \|z_{n_k}\|) + \delta'_{n_k} \right),$$

where z_{n_k} is uniformly bounded in Z and $\epsilon + k = \|A\overline{z} - Az_{n_k}\|$.

Since $z_{n_k} \to \overline{z}$ in Z and A is continuous, $\epsilon_k \to 0$ as $k \to \infty$. This implies that $A\overline{z} = \overline{u}$, and then, by the bijectivity of A, $\overline{z} = z$, i.e. $\overline{z}(s) = z(s)$ almost everywhere on $[a, b]$.

We show that $\overline{z}(s) = z(s)$ at every interior point $s \in (a, b)$. Assume that $\overline{z}(s_0) \ne \overline{z}(s_0)$. To be specific, suppose $\overline{z}(s_0) = z(s_0) + \epsilon$, $\epsilon > 0$. By the conditions of the theorem, $\overline{z}(s)$ is continuous on $[a, b]$, and hence also at s_0. Consequently, there is a $\delta > 0$ such that $|\overline{z}(s) - \overline{z}(s_0)| < \epsilon/2$ for all $s \in (s_0 - \delta, s_0 + \delta)$. Since $\overline{z}(s)$ in monotonically nonincreasing, for all $s \in (s_0 - \delta, s_0)$ we have $\overline{z}(s) - z(s) \ge \epsilon/2$. But then $\|\overline{z} - z\|_{L_p} > 0$, contradicting the fact that $\overline{z}(s) = z(s)$ almost everywhere on $[a, b]$.

Thus, $\overline{z}(s) = z(s)$ at each interior point $s \in (a, b)$. Since we can always extract a convergent subsequence from $\{z_n\}$, and since all these subsequences converge to the same function $\overline{z}(s)$, we can conclude that $z_n(s)$ itself converges to $\overline{z}(s)$ as $n \to \infty$, for each $s \in (a, b)$. We will now show that the sequence $z_n(s)$ converges uniformly on each subinterval $[\gamma, \sigma] \subset [a, b]$ to $\overline{z}(s)$ as $n \to \infty$. Fix an arbitrary $\epsilon > 0$. The uniform continuity of $\overline{z}(s)$ on $[\gamma, \sigma]$ implies that for all $\epsilon > 0$ there is a $\delta(\epsilon) > 0$ such that for any $s_1, s_2 \in [\gamma, \sigma]$ such that $|s_1 - s_2| < \delta(\epsilon)$ we have $|\overline{z}(s_1) - \overline{z}(s_2)| < \epsilon/4$.

Partition the interval $[\gamma, \sigma]$ by points $\gamma = s_1 < s_2 < \cdots < s_m = \sigma$ such that $|s_{i+1} - s_i| < \delta(\epsilon)$ $(i = 1, \ldots, m-1)$. Choose $N(\epsilon)$ such that for all s_i $(i = 1, \ldots, m)$ we have for $n > N(\epsilon)$:

$$|z_n(s_i) - \overline{z}(s_i)| < \frac{\epsilon}{4}.$$

The monotonicity of $\overline{z}(s)$ and $z_n(s)$ imply that for any $s \in [s_i, s_{i+1}]$ the graph of the function $z_n(s)$ belongs to a rectangle Π_i and, hence, the inequality $|z_n(s) - \overline{z}(s)| < 3\epsilon/4 < \epsilon$ holds, if only $n > N(\epsilon)$ (see Figure 2.1). $\square$

REMARKS. 1. The fact that, in general, $z_\eta(s)$ is not continuous is of some interest. In Theorem 2.1 the continuity of $\overline{z}(s)$ is important, and the approximation $z_\eta(s)$ may be a discontinuous monotone function. Nevertheless, $z_\eta(s) \to \overline{z}(s)$ as $\eta \to 0$, uniformly on every interval $[\gamma, \sigma] \subset (a, b)$.

2. Theorem 2.1 can be easily generalized to the case when the exact solution of (2.1) is a piecewise continuous function. In this case the sequence of approximations will converge to $\overline{z}(s)$, uniformly on every closed interval not containing a discontinuity point of $\overline{z}(s)$ or the points $s = a$, $s = b$.

Thus, if we know a priori that the exact solution $\overline{z}(s)$ of the ill-posed problem (2.1) is a monotone bounded function, we can take an arbitrary element $z_\eta \in Z{\downarrow}_C$ (η) as approximate solution. Here, $z_\eta \xrightarrow{\;L_p\;} \overline{z}$ as $\eta \to 0$. Moreover, if $\overline{z}(s)$ is a piecewise continuous function, the sequence of approximations converges to the exact solution, uniformly on every closed interval not containing a discontinuity point of $\overline{z}(s)$ or the points $s = a$, $s = b$.

Next to the a priori information that the exact solution $\overline{z}(s)$ of the ill-posed problem (2.1) be monotone, in a number of problems there are reasons to assume that $\overline{z}(s)$ is convex or monotone convex.

Consider the set $\breve{Z}_C$ of convex functions (upwards convex, to be specific) that are bounded above and below by constants C and 0, respectively. Also, let $\breve{Z}{\downarrow}_C$ be the set of (upwards) convex, nonincreasing, nonnegative functions that are bounded by the constant C ($\breve{Z}{\downarrow}_C = Z{\downarrow}_C \cap \breve{Z}_C$).

Next to $Z{\downarrow}_C$ (η) we will consider the sets $\breve{Z}_C(\eta)$ and $\breve{Z}{\downarrow}_C$ (η). Like $Z{\downarrow}_C$ (η), they are the sets of functions z_η from, respectively, $\breve{Z}_C$ and $\breve{Z}{\downarrow}_C$ such that

$$\|A_h z_\eta - u_\delta\| \leq \psi(h, \|z_\eta\|) + \delta.$$

As before, let $\eta_n \to 0$ as $n \to \infty$, and let $\overline{z}(s) \in \breve{Z}{\downarrow}_C$ (or $\breve{Z}_C$). We will assume that A is bijective from $\breve{Z}{\downarrow}_C$ $(\breve{Z}_C)$ onto $A\breve{Z}{\downarrow}_C$ $(A\breve{Z}_C)$.

THEOREM 2.2. *Let z_n be an arbitrary element in $\breve{Z}{\downarrow}_C$ (η_n) (respectively, $\breve{Z}_C(\eta_n)$), and let $[\gamma, \sigma]$ be a subinterval of (a, b). Then $z_n(s) \to \overline{z}(s)$ as $n \to \infty$, uniformly on $[\gamma, \sigma]$.*

For $\overline{z} \in \breve{Z}{\downarrow}_C$ (η_n) the truth of Theorem 2.2 follows from Theorem 2.1. For $\overline{z} \in \breve{Z}_C$, the theorem has been proved in [149].

REMARKS. 1. The continuity of $\overline{z}(s)$ at each interior point $s \in (a, b)$ follows from the fact that $\overline{z}(s) \in \breve{Z}_C$.

2. If $\overline{z}(s) \in \breve{Z}{\downarrow}_C$, Theorem 2.2 can be strengthened. Let z_n be an arbitrary element from $\breve{Z}{\downarrow}_C$ (η_n). Then $z_n(s) \to \overline{z}(s)$ as $n \to \infty$, uniformly on each interval $[a, b - \epsilon]$, $\epsilon > 0$.

To prove this fact it suffices to note that $\overline{z}(s) \overset{L_p}{=} \overline{\overline{z}}(s)$, $\overline{z}, \overline{\overline{z}} \in \breve{Z}{\downarrow}_C$, implies $\overline{z}(a) = \overline{\overline{z}}(a)$.

3. If $\overline{z}(s) \in \breve{Z}{\downarrow}_C$ (or $\breve{Z}_C$) we do not only have uniform convergence of the approximations, but under certain conditions even convergence of their derivatives [149].

Thus, for $\breve{Z}_C(s) \in \breve{Z}{\downarrow}_C$ (or $\breve{Z}_C$) we can take an arbitrary element from $\breve{Z}{\downarrow}_C$ (η) (respectively, $\breve{Z}_C(\eta)$) as approximate solution of (2.1).

The assertions proved in this section make it possible to use a minimization process for the discrepancy functional $\Phi(z) = \|A_h z - u_\delta\|^2$ on $Z{\downarrow}_C$, $\breve{Z}{\downarrow}_C$, or $\breve{Z}_C$, respectively, in order to find approximate solutions. Here, if the operators A and A_h are linear, it suffices to find an element z_η, in the sets mentioned above, for which $\Phi(z_\eta) \leq (hC_0 + \delta)^2$, where $C_0 = C\sqrt{b-a}$. For a linear problem (A and A_h linear), the problem of minimizing $\Phi(z)$ on $Z{\downarrow}_C$, $\check{Z}{\downarrow}_C$, or $\check{Z}_C$, is a convex programming problem.

3. Some theorems about convex polyhedra in $\mathbf{R}^n$

In this section we will consider the properties of piecewise approximations of the sets $Z{\downarrow}_C$, $\check{Z}{\downarrow}_C$, $\check{Z}_C$. These properties will be used in the sequel to construct algorithms for approximately solving (2.1) on the compact sets $Z{\downarrow}_C$, $\check{Z}{\downarrow}_C$, $\check{Z}_C$. For the numerical implementation of these algorithms we will use uniform grids for s on $[a,b]$.

When passing to finite difference approximations of the problem (2.1), the discrepancy functional $\Phi(z)$, which is defined on $Z{\downarrow}_C$, $\check{Z}{\downarrow}_C$, $\check{Z}_C$, is replaced by its finite difference approximation $\Phi(z)$, which is defined on all of $\mathbf{R}^n$. The sets $Z{\downarrow}_C$, $\check{Z}{\downarrow}_C$, $\check{Z}_C$ become the following sets of vectors in $\mathbf{R}^n$:[2]

$$M{\downarrow}_C = \left\{ z : z \in \mathbf{R}^n, \begin{array}{l} z_{i+1} - z_i \leq 0, \quad i = 1, \ldots, n-1, \\ 0 \leq z_i \leq C, \quad i = 1, \ldots, n \end{array} \right\}, \tag{2.2}$$

$$\check{M}_C = \left\{ z : z \in \mathbf{R}^n, \begin{array}{l} z_{i-1} - 2z_i + z_{i+1} \leq 0, \quad i = 2, \ldots, n-1, \\ 0 \leq z_i \leq C, \quad i = 1, \ldots, n \end{array} \right\}, \tag{2.3}$$

$$\check{M}{\downarrow}_C = \left\{ z : z \in \mathbf{R}^n, \begin{array}{l} z_{i-1} - 2z_i + z_{i+1} \leq 0, \quad i = 2, \ldots, n-1, \\ z_{i+1} - z_i \leq 0, \quad i = 1, \ldots, n-1, \\ 0 \leq z_i \leq C, \quad i = 1, \ldots, n \end{array} \right\}. \tag{2.4}$$

We will show that each of the sets $M{\downarrow}_C$, $\check{M}_C$, $\check{M}{\downarrow}_C$ is a closed convex bounded polyhedron in $\mathbf{R}^n$, and we will find the vertices of these polyhedra. To this end we need certain trivial facts regarding polyhedra in $\mathbf{R}^n$. We will use the following terminology.

We say that vectors $z_j \in \mathbf{R}^n$ ($j = 1, \ldots, m$) are *convexly independent* if none of them is a convex combination of the others.

A *convex polyhedron* in $\mathbf{R}^n$ is the convex hull of finitely many vectors $z_j \in \mathbf{R}^n$ ($j = 1, \ldots, l$).

A point z is called a *vertex* of a convex polyhedron $M = C\{z_j\}_{j=1}^l = \{x : x = \sum_{j=1}^l a_j z_j, a_j \geq 0, \sum_{j=1}^l a_j = 1\}$ if there is a vector $b \in \mathbf{R}^n$ such that for all $x \in M$ we have $(b, x - z) \geq 0$, with equality for $x = z$ only.

It is not difficult to prove the following trivial assertions [185].

LEMMA 2.2. *Let $M = C\{z_i\}_{i=0}^m$ be a convex polyhedron. Let t be a vertex of M. Then there is a p, $0 \leq p \leq m$, such that $z_p = t$.*

LEMMA 2.3. *Let $M = C\{z_i\}_{i=0}^m$ be a convex polyhedron. Then $z_0 \in M$ is a vertex of M if and only if z_0 is not a convex combination of the z_i ($i = 1, \ldots, m$).*

LEMMA 2.4. *Every convex polyhedron is the convex hull of its vertices.*

---

[2] Here, $z_i = z(s_i)$ are the values of the function $z(s)$ at the grid points.

This last lemma is a trivial consequence of Lemma 2.2 and Lemma 2.3. Thus, if we have to clarify whether or not a set of vectors z_i $(i = 0, \ldots, m)$ are the vertices of a convex polyhedron $M \subset \mathbf{R}^N$, we have to verify two conditions:

a) every $z \in M$ can be represented as a convex combination of the z_i $(i = 0, \ldots, m)$;

b) the vectors z_i $(i = 0, \ldots, m)$ are convexly independent.

THEOREM 2.3. *The set $M \downarrow_C$ is a convex polyhedron in $\mathbf{R}^n$, and its vertices $T^{(j)}$ $(j = 0, \ldots, n)$ can be written out explicitly:*

$$T^{(0)} = 0, \qquad T_i^{(j)} = \begin{cases} C, & i \le j, \\ 0, & i > j, \end{cases} \quad j = 1, \ldots, n. \qquad (2.5)$$

PROOF. We show that the $T^{(j)}$ $(j = 1, \ldots, n)$ are linearly independent. In fact, let

$$\sum_{j=1}^n c_j T^{(j)} = 0. \qquad (2.6)$$

We define a map δ_k from $\mathbf{R}^n$ into $\mathbf{R}$ as follows:

$$\delta_k u = u_{k+1} - u_k, \qquad k = 1, \ldots, n-1.$$

Applying δ_k to (2.6) gives:

$$0 = \delta_k \sum_{j=1}^n c_j T^{(j)} = \sum_{j=1}^n c_j \delta_k T^{(j)} = c_k \delta_k T^{(k)} = -C c_k.$$

This implies that $c_k = 0$ $(k = 1, \ldots, n-1)$.

Considering the nth component in (2.6) we see that $C c_n = 0$, i.e. $c_n = 0$. This proves that the $T^{(j)}$ $(j = 1, \ldots, n)$ are linearly independent.

We now show that any vector $M \downarrow_C$ can be represented as a convex combination of the $T^{(j)}$ $(j = 1, \ldots, n)$. Since the vectors $T^{(j)}$ form a basis of $\mathbf{R}^n$, there are numbers a_j such that

$$z = \sum_{j=1}^n a_j T^{(j)}. \qquad (2.7)$$

Applying the operator δ_k to this equality, we obtain

$$\delta_k z = a_k \delta_k T^{(k)} = a_k(-C), \qquad k = 1, \ldots, n-1.$$

Hence,

$$a_k = \frac{\delta_k z}{\delta_k T^{(k)}} = \frac{z_{k+1} - z_k}{-C}, \qquad k = 1, \ldots, n-1.$$

Since $z \in M \downarrow_C$, we have $\delta_k z \le 0$. Thus, $a_k \ge 0$ $(k = 1, \ldots, n-1)$. It is easily seen that $a_n = z_n / C \ge 0$.

For the first component (2.7) gives

$$\sum_{j=1}^n a_j = \frac{z_1}{C} \le 1.$$

Since $T^{(0)} = 0$,

$$z = \sum_{j=1}^{n} a_j T^{(j)} = \sum_{j=0}^{n} a_j T^{(j)},$$

with $a_j \geq 0$ $(j = 0, \ldots, n)$, $\sum_{j=0}^{n} a_j = 1$. Here, $a_0 = 1 - \sum_{j=1}^{n} a_j \geq 0$.

It remains to show that the vectors $T^{(j)}$ $(j = 0, \ldots, n)$ are convexly independent. It has been shown above that $T^{(j)}$ $(j = 1, \ldots, n)$ are linearly independent, so it is obvious that $T^{(0)} = 0$ cannot be a convex combination of these $T^{(j)}$. The assumption that one of the $T^{(j)}$ $(j = 1, \ldots, n)$ is a convex combination of the others leads also to linearly dependence of the vectors $T^{(j)}$. □

THEOREM 2.4. *The set $\check{M} \downarrow_C$ is a convex polyhedron in $\mathbf{R}^n$, and its vertices $T^{(j)}$ $(j = 0, \ldots, n)$ can be written out explicitly:*

$$T^{(0)} = \dot{0}, \qquad T_i^{(j)} = \begin{cases} C, & i \leq j, \\ \frac{n-i}{n-j} C, & i > j, \end{cases} \quad j = 1, \ldots, n. \tag{2.8}$$

PROOF. We show that the $T^{(j)}$ $(j = 1, \ldots, n)$ are linearly independent. We define a map $\delta_k^{(2)}$ from $\mathbf{R}^n$ into $\mathbf{R}$ as follows:

$$\delta_k^{(2)} u = u_{k-1} - 2u_k + u_{k+1}, \qquad k = 2, \ldots, n-1.$$

We assume that there are c_j such that

$$\sum_{j=1}^{n} c_j T^{(j)} = 0. \tag{2.9}$$

Applying δ_k to (2.9) gives:

$$\sum_{j=1}^{n} c_j \delta_k^{(2)} T^{(j)} = c_k \delta_k^{(2)} T^{(k)} = 0, \qquad k = 2, \ldots, n-1.$$

Since $\delta_k^{(2)} T^{(k)} < 0$, we have $c_k = 0$ $(k = 2, \ldots, n-1)$. By considering the nth component in (2.9), we find $c_n = 0$.

Applying the operator δ_1 to (2.9), we find $c_1 \delta_1 T^{(1)} = 0$. Since $\delta_1 T^{(1)} \neq 0$, we have $c_1 = 0$. This proves that the $T^{(j)}$ $(j = 1, \ldots, n)$ are linearly independent.

We will prove that any vector $z \in \check{M} \downarrow_C$ can be written as a convex combination of the $T^{(j)}$ $(j = 0, \ldots, n)$. Since the vectors $T^{(j)}$ $(j = 1, \ldots, n)$ are linearly independent, there are numbers a_j such that

$$z = \sum_{j=1}^{n} a_j T^{(j)}. \tag{2.10}$$

Applying the operator $\delta_k^{(2)}$ to (2.10), we obtain

$$\delta_k^{(2)} z = \sum_{j=1}^{n} a_j \delta_k^{(2)} T^{(j)} = a_k \delta_k^{(2)} T^{(k)}, \qquad k = 2, \ldots, n-1.$$

This implies that $a_k = \delta_k^{(2)} z / \delta_k^{(2)} T^{(k)}$ $(k = 2, \ldots, n - 1)$. Since $z \in \check{M} \downarrow_C$, we can easily be convinced of the fact that $a_k \geq 0$ $(k = 2, \ldots, n - 1)$. Using the nth component of (2.10), we see that $z_n = a_n C$, or $a_n = z_n / C \geq 0$.

Applying the operator δ_1 to (2.10) gives

$$\delta_1 z = \sum_{j=1}^{n} a_j \delta_1 T^{(j)} = a_1 \delta_1 T^{(1)}.$$

Hence, $a_1 = \delta_1 z / \delta_1 T \geq 0$. Thus, (2.10) determines the a_j $(j = 1, \ldots, n)$ uniquely, and all $a_j \geq 0$. Considering now the first component of (2.10), we find

$$z_1 = C \sum_{j=1}^{n} a_j,$$

or

$$\sum_{j=1}^{n} a_j = \frac{z_1}{C} \leq 1.$$

Consequently, any vector $z \in \check{M} \downarrow_C$ can be written as $z = \sum_{j=1}^{n} a_j T^{(j)}$ with $a_j \geq 0$ $(j = 0, \ldots, n)$, $\sum_{j=0}^{n} a_j = 1$. Here, $a_0 = 1 - \sum_{j=1}^{n} a_j \geq 0$.

It remains to prove the independence of the $T^{(j)}$ $(j = 0, \ldots, n)$. This is done completely similar as in the proof of Theorem 2.3. $\square$

THEOREM 2.5. *The set $\check{M}_C \subset \mathbf{R}^n$ is a convex polyhedron in $\mathbf{R}^n$, and its vertices $T^{(i,j)}$, $0 \leq i \leq j \leq n$, can be written out explicitly:*

$$T^{(0,0)} = 0, \qquad T_k^{(i,j)} = \begin{cases} \frac{k-1}{i-1}C, & k < i, \\ C, & i \leq k \leq j, \quad 1 \leq i \leq j \leq n. \\ \frac{n-k}{n-j}C, & k > j, \end{cases}$$

Note that if the polyhedra $M \downarrow_C$, $\check{M} \downarrow_C$ have $n + 1$ vertices, then $\check{M}_C$ has $1 + n(n+1)/2$ vertices.

THEOREM 2.6. *The set $\check{M}_C$ is contained in the convex polyhedron $\check{M}'_{2C}$, whose vertices $T^{(j)}$, $(j = 0, \ldots, n)$ can be written out explicitly:*

$$T^{(0)} = 0, \qquad T_i^{(j)} = \begin{cases} 2C\frac{i-1}{j-1}, & i < j, \\ 2C, & i = j, \quad j = 1, \ldots, n. \\ 2C\frac{n-i}{n-j}, & i > j, \end{cases} \qquad (2.11)$$

PROOF. Similarly as in the proofs of Theorem 2.3 and Theorem 2.4 we can easily verify that the $T^{(j)}$ $(j = 1, \ldots, n)$ are linearly independent.

Let $z \in \check{M}_C$ be arbitrary. We show that $z \in C\{T^{(j)}\}_{j=0}^{n}$. Since the $T^{(j)}$ $(j = 1, \ldots, n)$ are linearly independent, there are numbers a_j $(j = 1, \ldots, n)$ such that

$$z = \sum_{j=1}^{n} a_j T^{(j)}. \qquad (2.12)$$

Applying $\delta_k^{(2)}$ $(k = 2, \ldots, n-1)$ to (2.12) gives:

$$\delta_k^{(2)} z = \sum_{j=1}^{n} a_j \delta_k^{(2)} T^{(j)} = a_k \delta_k^{(2)} T^{(k)},$$

or

$$a_k = \frac{\delta_k^{(2)} z}{\delta_k^{(2)} T^{(k)}}, \qquad k = 2, \ldots, n-1.$$

Since $z \in \check{M}_C$, we see that $\delta_k^{(2)} z \leq 0$, and hence $a_k \geq 0$ $(k = 2, \ldots, n-1)$. Using the nth component of (2.12), we see that $a_n = z_n/(2C) \geq 0$. It remains to prove that $\sum_{j=1}^{n} a_j \leq 1$.

In fact, it is not difficult to see that

$$\sum_{i=1}^{n} T_i^{(j)} = \begin{cases} nC, & j = 1, n, \\ (n-1)C, & j = 2, \ldots, n-1. \end{cases}$$

Summing (2.12) componentwise, we find

$$\sum_{i=1}^{n} z_i = \sum_{j=1}^{n} a_j \sum_{i=1}^{n} T_i^{(j)} = a_1 nC + a_n nC + (n-1)C \sum_{j=2}^{n-1} a_j,$$

or

$$\sum_{i=1}^{n} z_i = a_1 C + a_n C + (n-1)C \sum_{j=1}^{n} a_j.$$

Using that $a_1 = z_1/(2C)$, $a_n = z_n/(2C)$, we find

$$C(n-1) \sum_{j=1}^{n} a_j = \frac{z_1}{2} + \frac{a_n}{2} + \sum_{j=2}^{n-1} z_j.$$

Since $z \in \check{M}_C$, we have $z_i \leq C$ $(i = 1, \ldots, n)$. But then $\sum_{j=1}^{n} a_j \leq 1$ by the previous equation.

Thus, for an arbitrary $z \in \check{M}_C$ there is a representation $z = \sum_{j=0}^{n} a_j T^{(j)}$ with $a_j \geq 0$, $\sum_{j=0}^{n} a_j = 1$. Here, $a_0 = 1 - \sum_{j=1}^{n} a_j \geq 0$.

Consequently, $\check{M}_C \subset \check{M}'_{2C} = C\{T^{(j)}\}_{j=0}^{n}$.

To finish the proof it suffices to verify the convex independence of the vectors $T^{(j)}$ $(j = 0, \ldots, n)$. This can be easily done, similarly as for Theorem 2.3 and Theorem 2.4. $\square$

REMARK. Theorem 2.6 implies that if the exact solution of the problem belongs to $\check{M}_C$, then we may look for an approximate solution in $\check{M}'_{2C}$. Indeed:

1) qualitative information regarding the upwards convexity of the 'function' is preserved, since $\check{M}'_{2C} \subset \check{M}_{2C}$;

2) The set of 'functions' in $\check{M}'_{2C}$ is bounded above by the constant $2C$, therefore all theorems regarding the convergence of approximations to the exact solution of the problem remain true when replacing $\check{M}_C$ by $\check{M}'_{2C}$.

Thus, we have studied in detail the structure of the polyhedra $M\!\downarrow_C$, $\check{M}_C$, $\check{M}\!\downarrow_C$. Each of them is a convex, bounded polyhedron in $\mathbf{R}^n$, whose vertices are known.

4. The solution of ill-posed problems on sets of convex functions

We return to Theorem 2.1 and Theorem 2.2 proved in §2.1. From the statement of these theorems it is clear that information regarding the uniform boundedness of the set of admissible functions is essential in the proofs of these theorems (the proofs rest on Helly's theorem). However, in the inverse problems of mathematical physics the typical situation is that we know a constant bounding the set of admissible functions from below (e.g., $C_2 = 0$, the admissible functions then being nonnegative), while the constant C is not known. We drop the uniform boundedness from above of the set of admissible functions. Let $Z\!\downarrow_C$, $\check{Z}$, $\check{Z}\!\downarrow_C$ be the set of monotone (nonincreasing), the set of convex (for the sake of being specific, upper convex), and the set of (upper) convex monotone (nonincreasing), bounded nonnegative functions, respectively. We will assume that the operator A is linear and bounded, and is known exactly, i.e. $h = 0$.[3] Similarly as in §2.2 we can introduce the sets $Z\!\downarrow_C (\delta)$, $\check{Z}(\delta)$, $\check{Z}\!\downarrow_C (\delta)$ of functions in $Z\!\downarrow_C$, $\check{Z}$, $\check{Z}\!\downarrow$, respectively, such that

$$\|Az - u_\delta\| \leq \delta.$$

Let δ_n be an arbitrary sequence such that $\delta_n \downarrow 0$ as $n \to \infty$.

Let z_n be an arbitrary element in $\check{Z}(\delta_n)$. We will show that the set of functions z_n is bounded, i.e. there is a constant $C > 0$ such that for any n we have $z_n(s) \leq C$ for all $s \in [a, b]$. This means that if the exact solution of the problem is a convex or monotonically convex function on $[a, b]$, then we may replace the bounded sets $\check{Z}_C(\eta)$ and $\check{Z}\!\downarrow_C (\eta)$ by $\check{Z}(\delta)$ and $\check{Z}\!\downarrow (\delta)$, respectively, when we are looking for an approximate solution.

It is not difficult to convince oneself of the truth of the following two lemmas. Let $f(x)$ be a nonnegative function defined and upper convex on the interval $[a, b]$.

LEMMA 2.5. *Suppose $f((a + b)/2) \leq 2$. Then $f(x) \leq 4$ for all $x \in [a, b]$.*

LEMMA 2.6. *Suppose that $f(x^*) > 4$ for some $x^* \in [a, b]$. Then $f((a + b)/2) > 2$.*

THEOREM 2.7. *Let $z_n \in \check{Z}(\delta_n)$ be arbitrary. Then there is a constant $C > 0$ such that for all n we have $z_n(s) \leq C$ for all $s \in [a, b]$.*

PROOF. By the linearity of A, the set $\check{Z}(\delta_n)$ is convex. Hence, at each point $s \in [a, b]$ the set of values of the functions $z(s) \in \check{Z}(\delta_n)$ is convex. In particular, this holds for $s = s^* = (a + b)/2$.

Now we assume the opposite, i.e. there is a subsequence of functions $z_{n_k}(s) \in \check{Z}(\delta_{n_k})$ and a sequence of points $s_k \in [a, b]$ such that $z_{n_k}(s_k) > k$. By Lemma 2.6, $z_{n_k}(s^*) > 2$ if only $k > 4$.

Note that $\overline{z}(s) \in \check{Z}(\delta_{n_k})$ and $\overline{z}(s^*) \leq 1$. Since the set of values at $s = s^*$ of the functions in $\check{Z}(\delta_{n_k})$ is convex, for each k there is a $z_k^0 \in \check{Z}(\delta_{n_k})$ such that $z_k^0(s^*) = 2$. By Lemma 2.5, the totality of functions $z_k^0(s)$ is uniformly bounded. By Helly's

---

[3] As before, we assume that $Z = L_p[a, b]$, $p > 1$.

theorem [95] we can choose a subsequence $z_{k_l}^0$ which pointwise converges to some function $\overline{\overline{z}} \in \check{Z}$. Clearly, $z_{k_l}^0 \xrightarrow{L_p} \overline{\overline{z}}$ as $l \to \infty$, since the $z_{k_l}^0$ are uniformly bounded.

Since A is continuous, $\|A\overline{\overline{z}} - \overline{u}\| = 0$, which implies that $\overline{\overline{z}} \stackrel{L_p}{=} \overline{z}$. On the other hand, $\overline{\overline{z}}(s^*) = 2$, while $\overline{z}(s^*) \leq 1$. Since both functions are continuous at s^*, this implies that $\overline{\overline{z}} \stackrel{L_p}{\neq} \overline{z}$. This is a contradiction. $\square$

Thus, if the exact solution of the ill-posed problem is convex, or monotone and convex, then to construct an approximate solution we do not need to know a constant bounding the exact solution from above. We will use this important fact in Chapter 3. If the exact solution is a monotone function on $[a, b]$, the similar assertion is, in general, not true.

5. Uniform convergence of approximate solutions of bounded variation

As before, we will consider the ill-posed problem

$$Az = u, \qquad z \in Z = L_p, \quad p > 1, \qquad u \in U.$$

Let A be a continuous operator from Z into U. We will assume that the exact solution $\overline{z}(s)$ of (2.1) is a continuous function of bounded variation on $[a, b]$. As usual, suppose we know, instead of A and $\overline{u}$, approximations of these: a continuous operator A_h and an element u_δ.

As before, we will construct algorithms that give uniform approximations to $\overline{z}(s)$. To this end we consider the set $V \subset L_p[a, b]$ of functions of bounded variation, and we will look for a 'smoothest' function among those functions in V that approximately satisfy (2.1). As measure of the smoothness we take the variation $V_a^b(z)$ of a function z. (This problem statement is convenient in certain inverse problems in spectroscopy, nuclear physics, and astrophysics, [71].) Since $V_a^b(z_1 - z_2) = 0$ implies that $z_1(s)$ and $z_2(s)$ differ by a constant, it is natural to fix the values of the functions at one endpoint of $[a, b]$. For the sake of being specific we will assume that we know the value of the exact solution $\overline{z}(s)$ at $s = a$. Without loss of generality we will assume that $\overline{z}(a) = 0$.

We let $V(\eta)$ be the set of elements $z \in V$ such that

1) $z(a) = 0$;
2) $\|A_h z - u_\delta\| < \psi(h, \|z\|) + \delta$.

We will assume that the operator A is bijective from V onto $AV \subset U$, i.e. $Az_1 \stackrel{U}{=} Az_2$ implies $z_1 \stackrel{L_p}{=} z_2$.

LEMMA 2.7. *Suppose $z_n \to z_0$ as $n \to \infty$ at every point $s \in [a, b]$. Moreover, suppose that $z_n \in V$ and that the $V_a^b(z_n)$ are totally bounded. Then $z_0 \in V$, and $V_a^b(z) \leq \liminf_{n \to \infty} V_a^b(z_n)$.*

PROOF. Take an arbitrary partition T of $[a, b]$ by points ξ_i $(i = 1, \ldots, m)$ and a sequence z_{n_k} such that $V_a^b(z_{n_k}) \to \liminf_{n \to \infty} V_a^b(z_n)$ as $k \to \infty$. Then

$$\sum_{i=1}^{m-1} |z_0(\xi_i) - z_0(\xi_{i+1})| = \lim_{k \to \infty} \sum_{i=1}^{m-1} |z_{n_k}(\xi_i) - z_{n_k}(\xi_{i+1})| \leq$$

$$\leq \lim_{k \to \infty} V_a^b(z_{n_k}) = \liminf_{n \to \infty} V_a^b(z_n). \qquad \square$$

LEMMA 2.8 ([185]). *Let* $\overline{z}(s), \overline{\overline{z}}(s) \in V$ *with* $\overline{z}(s)$ *continuous on* $[a, b]$. *Suppose that* $\overline{z}(s)$ *and* $\overline{\overline{z}}(s)$ *coincide almost everywhere on* $[a, b]$. *Then* $V_a^b(\overline{z}) \leq V_a^b(\overline{\overline{z}})$. *Unless* $\overline{\overline{z}}(s)$ *and* $\overline{z}(s)$ *are identical on* $[a, b]$, *we have* $V_a^b(\overline{z}) < V_a^b(\overline{\overline{z}})$.

THEOREM 2.8. *The functional* $V_a^b(z)$ *assumes its greatest lower bound on* $V(\eta)$.

PROOF. Let $z_n \in V(\eta)$ be a minimizing sequence:

$$\lim_{n \to \infty} V_a^b(z_n) = \inf_{z \in V(\eta)} V_a^b(z) = \overline{m}.$$

We will show that the set of functions z_n is uniformly bounded. Clearly, $V_a^b(z_n) \leq C$ for any n. On the other hand, the definition of $V_a^b(z)$ implies that for every $s \in [a, b]$,

$$|z_n(s)| = |z_n(s) - z_n(a)| \leq C.$$

The last relation proves that the sequence $z_n(s)$ is uniformly bounded.

By Helly's theorem [95] we can extract a subsequence $z_{n_k}(s)$ from $z_n(s)$ that converges at each point $s \in [a, b]$ to a function $\overline{z}(s)$. Lemma 2.7 implies that $\overline{z}(s) \in V$. Pointwise convergence and uniform boundedness imply L_p convergence. Thus, $z_{n_k} \overset{Z}{\longrightarrow} \overline{z}$ as $k \to \infty$. We are readily convinced of the fact that

$$\|A\overline{z} - u_\delta\| \leq \psi(h, \|\overline{z}\|) + \delta.$$

Hence, $\overline{z} \in V(\eta)$. On the other hand, by Lemma 2.7,

$$V_a^b(\overline{z}) \leq \liminf V_a^b(z_{n_k}) = \overline{m}. \qquad \square$$

REMARK. Theorem 2.8 has been proved under the assumption that all functions $z(s) \in V(\eta)$ have a fixed end: $z(a) = 0$. If A, acting from L_p into U, $p > 1$, is linear, continuous, and exactly known ($h = 0$), Theorem 2.8 can be proved also without this assumption. To see this we only have to prove the uniform boundedness (with respect to n) of the sequence $z_n(s)$. We will now prove that the sequence $z_n(s)$ is, in this case, uniformly bounded indeed.

Assume the opposite. Then there are a subsequence $z_{n_k}(s)$ and points $s_k \in [a, b]$ such that the sequence $z_{n_k}(s_k) = a_k$ becomes infinitely large. Clearly, $V_a^b(z_{n_k}) \leq C$ for all n_k. Therefore,

$$a_k - C \leq z_{n_k} \leq a_k + C$$

for all $s \in [a, b]$. Put $z^0(s) = 1$ for all $s \in [a, b]$, and let $y_k(s) = z_{n_k}(s) - z_{n_k}(a)$. Clearly, the norms of y_k are totally bounded. The linearity of A implies that $Az_{n_k} = z_{n_k}(a)Az^0 + Ay_k$. Note that $Az^0 \neq 0$, while the elements $\|Ay_k\|$ are totally bounded.

On the other hand, the number sequence $z_{n_k}(a)$ is unbounded, and hence the sequence of functionals

$$Az_{n_k} = z_{n_k}(a)Az^0 + Ay_k$$

is unbounded. This, however, contradicts the inequality $\|Az_{n_k} - u_\delta\| \leq \delta$. Thus, we have proved that the sequence of functions $z_n(s)$ is uniformly bounded. The remainder of the proof of Theorem 2.8 can be given without changes.

Let now $z_\eta \in V(\eta)$ be an arbitrary element for which

$$V_a^b(z_\eta) = \inf_{z \in V(\eta)} V_a^b(z).$$

THEOREM 2.9 ([66]). *Let $\overline{z} \in C[a,b] \cap V$. Then $z_\eta \to \overline{z}$ uniformly on $[a,b]$ as $\eta \to 0$.*

PROOF. Let $\eta_n \to 0$ as $n \to \infty$. We will first show that $z_{\eta_n}(s) \to \overline{z}(s)$ at every point $s \in [a,b]$. Note that since $\overline{z} \in V(\eta)$ for all $\eta > 0$, we have $V_a^b(z_{\eta_n}) \leq V_a^b(\overline{z})$. Similarly as in the proof of Theorem 2.8 we can prove that the set of functions $z_{\eta_n}(s)$ is uniformly bounded. So, we can extract a subsequence $z_k^0 = z_{\eta_{n_k}}$ from it which converges pointwise to an element $\overline{\overline{z}} \in V$, and hence converges in the metric of Z. By limit transition we can readily see that $A\overline{\overline{z}} = A\overline{z}$, and hence $\overline{\overline{z}} \overset{Z}{=} \overline{z}$, i.e. $\overline{\overline{z}}(s)$ and $\overline{z}(s)$ coincide almost everywhere on $[a,b]$. Since $V_a^b(z_k^0) \leq V_a^b(\overline{z})$, Lemma 2.7 implies

$$V_a^b(\overline{\overline{z}}) \leq \liminf_{k \to \infty} V_a^b(z_k^0) \leq V_a^b(\overline{z}),$$

i.e.,

$$V_a^b(\overline{\overline{z}}) \leq V_a^b(\overline{z}).$$

By Lemma 2.8, $\overline{\overline{z}}(s) = \overline{z}(s)$ everywhere on $[a,b]$. Thus, the sequence $z_k^0(s)$ converges to $\overline{z}(s)$ at every point $s \in [a,b]$, as $k \to \infty$. The proof that the sequence $z_{\eta_n}(s)$ converges pointwise to $\overline{z}(s)$ presents no difficulties now.

We will prove that the sequence $z_{\eta_n}(s)$ converges uniformly on $[a,b]$ to $\overline{z}(s)$, as $n \to \infty$. Assume the opposite. Then there are a subsequence $z_{\eta_{n_k}} = z_k^0$ and numbers $\xi_k \in [a,b]$, $\epsilon > 0$, such that

$$|z_k^0(\xi_k) - \overline{z}(\xi_k)| > \epsilon.$$

Consider now a partition T by points x_p, $p = 1, \ldots, m$, such that

$$\sum_{p=1}^{m-1} |\overline{z}(x_p) - \overline{z}(x_{p+1})| > V_a^b(\overline{z}) - \frac{\epsilon}{2}.$$

It is clear that, since $\overline{z}(s)$ is uniformly continuous on $[a,b]$, we can choose T such that for any $x', x'' \in [x_p, x_{p+1}]$, $p = 1, \ldots, m-1$, we have

$$|\overline{z}(x') - \overline{z}(x'')| < \frac{\epsilon}{6}.$$

Since there is pointwise convergence, proved above, there is an $N(\epsilon)$ such that for all $k > N(\epsilon)$,

$$|z_k^0(x_p) - \overline{z}(x_p)| < \frac{\epsilon}{4(m-1)}, \qquad p = 1, \ldots, m.$$

Choose an arbitrary $k > N(\epsilon)$ and consider z_k^0. We will obtain a lower bound for its variation on $[a, b]$. Let $\xi_k \in [x_l, x_{l+1}]$. Then for $k > N(\epsilon)$ we have

$$V_a^b(z_k^0) \geq \sum_{\substack{p=1 \\ p \neq l}}^{m-1} |z_k^0(x_p) - z_k^0(x_{p+1})| + |z_k^0(x_l) - z_k^0(\xi_k)| + |z_k^0(x_{l+1}) - z_k^0(\xi_k)| \geq$$

$$\geq \sum_{\substack{p=1 \\ p \neq l}}^{m-1} |\overline{z}(x_p) - \overline{z}(x_{p+1})| - \frac{2\epsilon(m-2)}{4(m-1)} - \frac{2\epsilon}{4(m-1)} +$$

$$+ |\overline{z}(x_l) - z_k^0(\xi_k)| + |\overline{z}(x_{l+1}) - z_k^0(\xi_k)| \geq$$

$$\geq \sum_{p=1}^{m-1} |\overline{z}(x_p) - \overline{z}(x_{p+1})| + 2|z_k^0(\xi_k) - \overline{z}(\xi_k)| - \frac{\epsilon}{2} - \frac{3\epsilon}{6} \geq$$

$$\geq V_a^b(\overline{z}) + 2\epsilon - \epsilon - \frac{\epsilon}{2} = V_a^b(\overline{z}) + \frac{\epsilon}{2}.$$

The last inequality contradicts the definition of the element $z_k^0(s) = z_{\eta_{m_k}}(s)$ as an element of minimal variation on the set. $\square$

REMARKS. 1. Thus, if we know a priori that the exact solution of the ill-posed problem (2.1) is a continuous function of bounded variation, then we can take as approximate solution any element z_η realizing the minimum of $V_a^b(z)$ on $V(\eta)$. Moreover, uniform convergence of the sequence of approximate solutions to the exact solution of (2.1) is guaranteed. In Chapter 3 we will consider algorithms solving this problem.

2. If A is linear, continuous and exactly known ($h = 0$) on L_p, then Theorem 2.9 remains true if we drop the assumption that all functions $z(s) \in V(\eta)$ have a fixed end. The proof of the uniform boundedness of the functions $z_{\eta_n}(s)$ can be given as in the remark to Theorem 2.8. The remainder of the proof is the same as that of Theorem 2.9.

CHAPTER 3

Algorithms for the approximate solution of ill-posed problems on special sets

In Chapter 2 we have succeeded in solving, in a number of cases, the first problem posed to us: starting from qualitative information regarding the unknown solution, how to find the compact set of well-posedness M containing the exact solution. It was shown that this can be readily done if the exact solution of the problem belongs to $Z\downarrow_C$, $\check{Z}_C$, $\check{Z}\downarrow_C$. A uniform approximation to the exact solution of the problem can be constructed if the exact solution is a continuous function of bounded variation. We now turn to the second problem: how to construct an efficient numerical algorithm for solving ill-posed problems on the sets listed above?

1. Application of the conditional gradient method for solving problems on special sets

We will first treat the solution of the ill-posed problem

$$Az = u, \qquad z \in M \subset Z = L_2, \quad u \in U, \tag{3.1}$$

under the condition that the operator A is continuous, linear and exactly known $(h = 0)$ from Z into U. We take for M one of the sets $Z\downarrow_C$, $\check{Z}_C$, or $\check{Z}\downarrow_C$. We will assume that U is a Hilbert space. It has been shown in Chapter 2 that for $\bar{z} \in Z\downarrow_C$, $\check{Z}_C$, or $\check{Z}\downarrow_C$ we can take as approximate solution of the problem (3.1) any element z_δ in $Z\downarrow_C$, $\check{Z}_C$, $\check{Z}\downarrow_C$, respectively, for which $\|Az_\delta - u_\delta\| \leq \delta$.

Each of $Z\downarrow_C$, $\check{Z}_C$, $\check{Z}\downarrow_C$ is a bounded closed convex set Z. Consider the functional

$$\Phi(z) = \|Az - u\|^2, \tag{3.2}$$

which is defined for all z in $Z\downarrow_C$, $\check{Z}_C$, $\check{Z}\downarrow_C$, respectively. Since A is linear, $\Phi(z)$ is a quadratic function of z. Clearly, $\Phi(z)$ is convex and differentiable, while its Fréchet derivative [95] is equal to

$$\Phi'(z) = 2(A^*Az - A^*u_\delta). \tag{3.3}$$

Here, $A^*: U \to Z$ is the adjoint of A. Note that

$$\|\Phi'(z_1) - \Phi'(z_2)\| = 2\|A^*A(z_1 - z_2)\| \leq 2\|A\|^2\|z_1 - z_2\|.$$

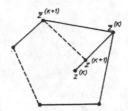

FIGURE 3.1. Aid in the derivation of (3.4).

Thus, the Fréchet derivative of $\Phi(z)$ satisfies the Lipschitz condition with constant $L = 2\|A\|^2$.

In our case the problem of finding an approximate solution can be solved by minimizing $\Phi(z)$ on $Z\downarrow_C$, $\check{Z}_C$, $\check{Z}\downarrow_C$, respectively. Here it is not necessary to find the minimum of $\Phi(z)$ on these sets, but it suffices to find an element z_δ in these sets such that $\Phi(z_\delta) \leq \delta^2$. Thus, to find an approximate solution of (3.1) on $Z\downarrow_C$, $\check{Z}_C$, $\check{Z}\downarrow_C$ we have to study the construction of a sequence minimizing some convex differentiable functional on a closed convex bounded set in a Hilbert space.

If, as in our case, the Fréchet derivative of the functional satisfies a Lipschitz condition, then to solve the above problem we may use, e.g., the conditional gradient method [32], [33], [76], [208].

In the conditional gradient method one constructs, next to a minimizing sequence $z^{(k)}$, an auxiliary sequence $\overline{z}^{(k)}$ as follows. Start with an arbitrary admissible point $z^{(0)} \in M$.[1] Suppose $z^{(k)}$ has been constructed. Then $\overline{z}^{(k)}$ is a solution of the problem

$$\left(\Phi'(z^{(k)}, \overline{z}^{(k)}\right) = \min_{z \in M} \left(\Phi'(z^{(k)}, z\right). \tag{3.4}$$

This problem is solvable, i.e. there is a (in general, nonunique) point $\overline{z}^{(k)}$ at which the linear functional $\left(\Phi'(z^{(k)}, z\right)$ assumes its minimal value on M [27]. Clearly, $\overline{z}^{(k)}$ belongs to the boundary of M (see Figure 3.1). Note that the problem (3.4) can be solved very simply in case M is a bounded closed convex polyhedron in $\mathbf{R}^n$. Then (3.4) is a linear programming problem, which can be solved by the simplex method [93] or, if the vertices of M are known, by simply checking all vertices.

Suppose we have found $\overline{z}^{(k)}$. Then $z^{(k+1)}$ is constructed in accordance with

$$z^{(k+1)} = z^{(k)} + \lambda_k(\overline{z}^{(k)} - z^{(k)}),$$

where $\lambda_k \in [0, 1]$ is the solution of the one-dimensional minimization problem

$$\Phi(z^{(k+1)}) = \Phi\left(z^{(k)} + \lambda_k(\overline{z}^{(k)} - z^{(k)})\right) = \min_{\lambda \in [0,1]} \Phi\left(z^{(k)} + \lambda(\overline{z}^{(k)} - z^{(k)})\right). \tag{3.5}$$

The latter problem comes down to minimizing $\Phi(z)$ on the segment $[z^{(k)}, \overline{z}^{(k)}]$. Since M is convex, $z^{(k+1)} \in M$. Thus, starting the iteration process with $z^{(0)} \in M$, the minimization process does not lead outside the boundary of M.

---

[1] An admissible point is any element of M. Since all sets $Z\downarrow_C$, $\check{Z}_C$, $\check{Z}\downarrow_C$ contain 0, we can take $z^{(0)} = 0$.

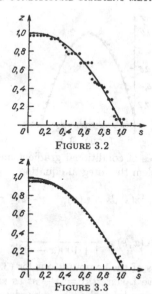

FIGURE 3.2

FIGURE 3.3

If $\Phi(z)$ is a quadratic function (this is true if A is linear), then (3.5) is a trivial problem: find the minimum of the parabola (with respect to $\lambda \in [0,1]$). If A is linear, the sequence $z^{(k)}$ thus constructed is minimizing for $\Phi(z)$ on M [27].

Thus, since in our case $\Phi(z)$ is a quadratic function and (3.5) can be trivially solved, to construct efficient algorithms for the approximate solution of the ill-posed problem (3.1) on $Z \downarrow_C$, $\check{Z}_C$, $\check{Z} \downarrow_C$ it suffices to study the efficient solution of the problem (3.4). In Chapter 2, §3 we have given a simple computation of the vertices of the polyhedra $M\downarrow_C$, $\check{M}\downarrow_C$, $\check{M}_C$, $\check{M}'_{2C}$ into which the sets $Z\downarrow_C$, $\check{Z}\downarrow_C$, $\check{Z}_C$ transform under finite-difference approximation. The problem (3.4) can now simply be solved by checking the vertices of these polyhedra. Here, in the solution of (3.1) on the set of convex functions $\check{Z}_C$, the polyhedron $\check{M}_C$ can be replaced by the polyhedron $\check{M}'_{2C}$ (cf. the remark to Chapter 2, Theorem 2.6). The latter polyhedron has $n+1$ vertices and in solving (3.4) one does not encounter any problems. So, the linear programming problem (3.4) can be solved at each iteration using n steps by checking the vertices of the polyhedra listed above.

REMARK. Consider the case when A is given approximately. Suppose the exact solution of (3.1) belongs to one of the sets $Z\downarrow_C$, $\check{Z}\downarrow_C$, $\check{Z}_C$. In this case (see Chapter 2), as approximate solution we can take an arbitrary element from $Z\downarrow_C$, $\check{Z}\downarrow_C$, $\check{Z}_C$, respectively, satisfying $\|A_h z_\eta - u_\delta\| \le \psi(h, \|z_\eta\|) + \delta$ (if A, A_h are linear, then it has to satisfy $\|A_h z_\eta - u_\delta\| \le h\|z_\eta\| + \delta$). Clearly, in this inequality we can replace the norm of z_η by its least upper bound $C_0 = \sup \|z\|$ on $Z\downarrow_C$, $\check{Z}\downarrow_C$, $\check{Z}_C$, respectively. For example, we can take $C_0 = C(b-a)^{1/2}$. Now we can construct algorithms for approximately solving the ill-posed problem (3.1) on the sets $Z\downarrow_C$, $\check{Z}\downarrow_C$, $\check{Z}_C$ in the same way as when the operator is exactly known.

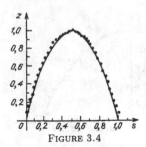

FIGURE 3.4

We illustrate the possibilities of conditional gradient methods on special sets. As before, we take as model problem the integral equation

$$\int_0^1 K(x,s)z(s)\,ds = u(x), \qquad -2 \le x \le 2, \qquad (3.6)$$

with kernel

$$K(x,s) = \frac{1}{1 + 100(x-s)^2}.$$

Figure 3.2 contains the results of the computations for (3.6) on the set of monotone decreasing functions. We have taken $\overline{z}(s) = 1 - s^2$ as exact solution of (3.6), the righthand side of (3.6) has been perturbed by a random error which is uniformly distributed on $[-\delta_0, \delta_0]$, where δ_0 is 1% of the maximum value of the righthand side. Here, the mean-square error corresponds to $\delta^2 = 7.36 \cdot 10^{-6}$. The exact solution and the approximate solution found after 400 iterations are depicted in Figure 3.2.[2] The approximate solution found corresponds to the following value of the discrepancy functional: $\Phi(z) = 7.97 \cdot 10^{-6}$. The subsequent application of the conditional gradient method changes the discrepancy and solution only slightly. We have used the initial approximation $z^{(0)}(s) = 0$.

Figure 3.3 contains the results for the same model equation (3.6) on the set of monotone convex functions. We have taken $\overline{z}(s) = 1 - s^2$ as exact solution. The level of perturbation of the initial data is 3% of the maximum of the righthand side. The approximate solution found corresponds to the following value of the discrepancy functional: $\Phi(z) = 6.62 \cdot 10^{-5}$ and has been obtained after 50 iterations of the conditional gradient method. The initial approximation is $z^{(0)}(s) = 0$.

The potential of the conditional gradient method is demonstrated in Figure 3.4. Here, (3.6) is solved on the set of convex functions. Th e exact solution is $\overline{z}(s) = 4s(1-s)$, and the righthand side is perturbed in such a way that $\delta^2 = 8.22 \cdot 10^{-6}$. The approximate solution, obtained after 800 iterations of the conditional gradient method, corresponds to the discrepancy value $9.31 \cdot 10^{-6}$, which is about 1% of the maximum of the righthand side.

The conditional gradient method can also be used to construct approximate solutions of (3.1) on the set V of functions of bounded variation. In Chapter 2, §5 we have proved that if the exact solution is a continuous function of bounded variation, then

---

[2] In Figure 3.2 to Figure 3.12, the dots correspond to the approximate solution, and the continuous line to the exact solution.

as approximate solution we may take an arbitrary element z_η realizing the minimum of $V_a^b(z)$ on $V(\eta)$. For simplicity we will assume that A is linear and exactly known ($h = 0$). Then we can take as approximate solution of (3.1) an element z_δ such that

$$V_a^b(z_d) = \min_{z \in V(\eta)} V_a^b(z) = \overline{m}.$$

We will solve this problem as follows. Let C be an arbitrary constant, $C \geq \overline{m}$. We pose the following problem. Find, in the set $V_C = \{z\colon V_a^b(z) \leq C\}$ of functions whose variations do not exceed C, an element z_C such that $\|Az_C - u_\delta\| \leq \delta$.[3] This problem can be solved by minimizing the discrepancy functional $\Phi(z) = \|Az_C - u_\delta\|^2$ on the set of functions $z \in V_C$. We have to proceed with this minimization until $\Phi(z)$ becomes less than or equal to δ^2. Below we will give efficient numerical algorithms for solving ill-posed problems on V_C.

It now remains to find the least C for which the condition

$$\min_{z \in V_C} \Phi(z) \leq \delta^2$$

holds, i.e. the problem of finding z_C is solvable. This problem can be approximately solved in a simple manner (reducing C) on a chosen grid of values of the parameter C.

Thus, we have arrived at the following problem: Construct a sequence minimizing the quadratic functional $\Phi(z)$ on the set of functions of variation not exceeding a given constant C. After transition to a finite-difference approximation there arises the question: Construct a sequence of vectors $z^{(k)} \in \mathbf{R}^n$ minimizing a quadratic functional $\varphi(z)$ on the set M_{V_C} of vectors $z \in \mathbf{R}^n$ whose components satisfy the inequality

$$|z_2 - z_1| + |z_3 - z_2| + \cdots + |z_n - z_{n-1}| \leq C.$$

For simplicity we assume that one boundary value, $\overline{z}(a)$ or $\overline{z}(b)$, is known. Without loss of generality we may assume that $\overline{z}(b) = 0$. It is then natural to put $z_n = 0$.

We will consider in more detail the question of the structure of the set M_{V_C} of vectors $z \in \mathbf{R}^n$ whose components satisfy the relations

$$|z_2 - z_1| + |z_3 - z_2| + \cdots + |z_n - z_{n-1}| \leq C,$$
$$z_n = 0.$$

Similarly as in Chapter 2, §3 we will show that M_{V_C} is a convex polyhedron in $\mathbf{R}^n$ with $2(n-1)$ vertices. But first we will prove a lemma. Suppose we are given vectors $T^{(j)}$ $(j = -(n-1), \ldots, -1, 1, \ldots, n-1)$ in $\mathbf{R}^n$ with components

$$T_i^{(j)} = \begin{cases} C, & i \leq j, \\ 0, & i > j, \end{cases} \qquad j = 1, \ldots, n-1,$$

$$T^{(-j)} = -T^{(j)}, \qquad j = 1, \ldots, n-1. \tag{3.7}$$

---

[3] This problem is of independent interest, since V_C is compact in L_p, $p > 1$. Thus, the approximate solution obtained in the well-posedness set V_C converges to the exact solution $\overline{z}$ in L_p, $p > 1$.

LEMMA 3.1. *The vectors $T^{(j)}$ $(j = -(n-1),\ldots,-1,1,\ldots,n-1)$ are convexly independent.*

PROOF. Assume the opposite. Suppose that for some k (without loss of generality we may assume $k > 0$)

$$T^{(k)} = \sum_{\substack{j=-(n-1)\\j\neq k,0}}^{n-1} a_j T^{(j)},$$

$$a_j \geq 0, \qquad \sum_{\substack{j=-(n-1)\\j\neq k,0}}^{n-1} a_j = 1. \tag{3.8}$$

Let δ_k be the operator from $\mathbf{R}^n$ into $\mathbf{R}$ introduced in Chapter 2, §3: $\delta_k u = u_{k+1} - u_k$ $(k = 1,\ldots,n-1)$. Letting δ_k act on (3.8) gives

$$\delta_k T^{(k)} = \sum_{\substack{j=-(n-1)\\j\neq k,0}}^{n-1} a_j \delta_k T^{(j)} = a_{-k}\delta_k T^{(-k)}.$$

Since $\delta_k T^{(k)} = -\delta_k T^{(-k)} \neq 0$, the last equation implies that $a_{-k} < 0$. This contradiction proves the lemma. $\square$

THEOREM 3.1. *The set M_{V_C} of vectors is a convex polyhedron in $\mathbf{R}^n$ whose vertices $T^{(j)}$ $(j = -(n-1),\ldots,-1,1,\ldots,n-1)$ have the form (3.7).*

PROOF. Since the convex independence of the $T^{(j)}$ has been shown in Lemma 3.1, it suffices to prove that an arbitrary vector $z \in M_{V_C}$ can be written as a convex combination of the $T^{(j)}$, i.e.

$$z = \sum_{\substack{j=-(n-1)\\j\neq 0}}^{n-1} a_j T^{(j)},$$

with

$$a_j \geq 0, \qquad \sum_{\substack{j=-(n-1)\\j\neq 0}}^{n-1} a_j = 1.$$

We will prove that such a_j exist. Applying δ_k $(k = 1,\ldots,n-1)$ to the expansion of z with respect to the $T^{(j)}$, we obtain

$$\delta_k z = \sum_{\substack{j=-(n-1)\\j\neq 0}}^{n-1} a_j \delta_k T^{(j)} = a_k \delta_k t^{(k)} + a_{-k}\delta_k T^{(-k)}.$$

We choose a_k in accordance with the following rule: If $\delta_k z \geq 0$, then

$$a_k = 0, \qquad a_{-k} = \frac{\delta_k z}{\delta_k T^{(-k)}} = \frac{\delta_k z}{C} \geq 0, \qquad k = 1,\ldots,n-1;$$

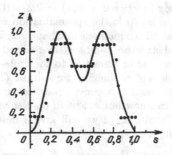

FIGURE 3.5

if, however, $\delta_k z \leq 0$, then

$$a_k = \frac{\delta_k z}{\delta_k T^{(k)}} = \frac{\delta_k z}{-C} \geq 0, \qquad a_{-k} = 0, \qquad k = 1, \ldots, n-1.$$

It can be readily seen that the a_j thus constructed are the coefficients of the expansion of z with respect to the $T^{(j)}$. Moreover, if $z \in M_{V_C}$, then

$$a_j > 0, \qquad \sum_{\substack{j=-(n-1) \\ j \neq 0}}^{n-1} a_j = \frac{|z_2 - z_1| + \cdots + |z_n - z_{n-1}|}{C} \leq 1.$$

If this sum is 1, the theorem has been proved. Suppose

$$\sum_{\substack{j=-(n-1) \\ j \neq 0}}^{n-1} a_j = \gamma < 1.$$

In this case it suffices to take

$$a_1' = a_1 + \frac{1-\gamma}{2}, \qquad a_{-1}' = a_{-1} + \frac{1-\gamma}{2}$$

instead of a_1 and a_{-1}. $\square$

Now, the problem of constructing a minimizing sequence for the functional $\Phi(z)$ on the set V_C of functions whose variations do not exceed C can be easily solved by the conditional gradient method. (As already noted, this problem is of independent interest for applications.)

In Figure 3.5 we have shown the potential of the conditional gradient method for solving problems on the set of functions whose variations do not exceed a given constant.

The exact solution of (3.6) is

$$\bar{z}(s) = \left(e^{-(s-0.3)^2/0.03} + e^{-(s-0.7)^2/0.03} \right) 0.9550408 - 0.052130913.$$

We perturb the righthand side of (3.6) by a uniformly distributed random error such that $\delta^2 = 5.56 \cdot 10^{-6}$. The approximate solution $z(s)$ is obtained after 800 iterations.

The value of the discrepancy functional is $\Phi(z) = 2.27 \cdot 10^{-5}$, which is 1.5% of the maximum of the righthand side. As initial approximation we have used $z^{(0)} = 0$.

Thus, we have proved that all the methods in Chapter 2, §2, §5 for solving ill-posed problems on sets of a special structure can be easily realized by the conditional gradient method, since in each case, after transition to the finite-difference approximation we have arrived at the problem of minimizing a quadratic (A being linear) functional on a set which can be represented as a convex polyhedron in $\mathbf{R}^n$ whose vertices we can express explicitly. We can now implement in an elementary way algorithms for solving the problem. In Chapter 4, §2 we will give a description of the programs implementing these algorithms.

REMARKS. 1. Note that even if the operator A is known exactly, when passing to the finite-difference approximation we replace it by a finite-dimensional operator A_h. In this case we have to control the approximation error (the choice of the amount of grid nodes) such that the inequality $hC_0 \ll \delta$ is satisfied.

2. The complications arising in the attempt to transfer the results of this Chapter to the case of a nonlinear operator A are well-known and are related with the fact that the functional $\Phi(z) = \|Az - u_\delta\|^2$ need not be convex. Therefore it is impossible to guarantee the convergence of the iteration algorithms with an arbitrary admissible initial approximation.

2. Application of the method of projection of conjugate gradients to the solution of ill-posed problems on sets of special structure

In the previous section we have proposed algorithms for the approximate solution of ill-posed problems on special sets, based on the conditional gradient method. Such methods have been widely applied for the solution of a large number of practical problems. However, as practice has revealed, the iteration algorithms given are in fact efficient only when the initial information is specified with a relatively small degree of exactness (approximately in the order of a few percents). If smaller values of the discrepancy functional $\Phi(z)$ are required (a larger degree of exactness of specifying the initial data), the computing time increases sharply. On the other hand, in certain inverse problems of physics [71] the degree of exactness of specifying the initial data is presently of the order of parts of a percent, and this is not yet the ultimate order. Therefore it became necessary to create superefficient algorithms for solving ill-posed problems with constraints. A number of interesting algorithms for solving ill-posed problems on special sets can be found in [48], [141], [149], [150].

In §2, §3 of the present Chapter we will propose efficient algorithms for solving ill-posed problems on special sets, solving the problem of constructing an approximate solution in finitely many steps.

Suppose the exact solution of (3.1) belongs to one of the sets $Z\!\downarrow_C$, $\check{Z}_C$, or $\check{Z}\!\downarrow_C$. For the sake of simplicity we will assume that the operator A is exactly known and linear. We have already convinced ourselves of the fact that we may take as approximate solution an arbitrary element z_δ belonging to $Z\!\downarrow_C$, $\check{Z}_C$, or $\check{Z}\!\downarrow_C$, respectively, and such that $\|Az - u_\delta\| \leq \delta$. As has been proved in Chapter 2, if $\overline{z}(s) \in \check{Z}_C$ or $\check{Z}\!\downarrow_C$, then we may look for an approximate solution in $\check{Z}$ or $\check{Z}\!\downarrow$, respectively, i.e. we can drop the uniform boundedness of the functions by a constant C (see Chapter 2, Theorem 2.7). Under finite-difference approximation, the sets $Z\!\downarrow_C$, $\check{Z}$, $\check{Z}\!\downarrow$ become the convex sets

$$M\!\downarrow_C = \left\{ z : \begin{array}{ll} z_{i+1} - z_i \leq 0, & i = 1, \ldots, n-1 \\ 0 \leq z_i \leq C, & i = 1, \ldots, n \end{array} \right\}, \tag{3.9}$$

$$\check{M} = \left\{ z : \begin{array}{ll} z_{i-1} - 2z_i + z_{i+1} \leq 0, & i = 2, \ldots, n-1 \\ z_i \geq 0, & i = 1, \ldots, n \end{array} \right\}, \tag{3.10}$$

$$\check{M}\!\downarrow = \left\{ z : \begin{array}{ll} z_{i-1} - 2z_i + z_{i+1} \leq 0, & i = 2, \ldots, n-1 \\ z_{i+1} - z_i \leq 0, & i = 1, \ldots, n-1 \\ z_i \geq 0, & i = 1, \ldots, n \end{array} \right\}. \tag{3.11}$$

The functional $\Phi(z) = \|Az - u_\delta\|^2$ becomes the quadratic function $\varphi(z)$. Thus, we have passed to the following problem: Construct a minimizing sequence for the functional $\Phi(z)$ on one of the sets (3.9)–(3.11).

Let Y denote one of the sets $M\!\downarrow_C$, $\check{M}$, or $\check{M}\!\downarrow$. Note that any constraint in (3.9)–(3.11) can be written as

$$Fz \leq g, \tag{3.12}$$

where F is a matrix of dimensions $m_0 \times n$, m_0 being the amount of constraints defining the set, and g is a vector of length m_0. An inequality is to be understood as componentwise inequalities.

If z is on the boundary of a set, then one or several inequalities in (3.12) may turn into an equality. We will call the set of indices for which at a point z the equality

$$\sum_{j=1}^{n} F_{ij} z_j = g_i$$

holds, the *set of active constraints* (at z), and we denote it by $I(z)$. By $\dim I$ we denote the number of elements in I. The *matrix of active elements* is the matrix F_I of dimensions $\dim I \times n$ whose rows are the rows of F that have row-number in $I(z)$.

We write the function $\varphi(z)$ as

$$\varphi(z) = (z, Qz) + (d, z) + e. \tag{3.13}$$

The simplest way to state the method of projection of conjugate gradients for minimizing the function (3.13) under the constraints (3.12) is in algorithmic form [145], [146].

Step 1. The minimization starts with an arbitrary admissible point $z^{(0)} \in Y$ and proceeds in $\mathbf{R}^n$ by the method of conjugate gradients. The number of active constraints is put equal to zero: $m = 0$. The iteration counter of the method of projection of conjugate gradients without changing the matrix of active constraints is put equal to zero: $k = 0$.

Step 1.1. Since the method of conjugate gradients finds the minimum in $\mathbf{R}^n$ in n steps, if $k = n$ we have found a solution. In this case, go to Step 6.

Step 1.2. Compute the direction of descent $p^{(k)}$: if $k = 0$, then

$$p^{(k)} = -\operatorname{grad}\varphi(z^{(k)}),$$

otherwise

$$p^{(k)} = -\operatorname{grad}\varphi|_{z=z^{(k)}} + \frac{\|\operatorname{grad}\varphi(z^{(k)})\|^2}{\|\operatorname{grad}\varphi(z^{(k-1)})\|^2}p^{(k-1)}.$$

Step 1.3. Compute a_k, the optimal stepsize along this direction, by the formula

$$a_k = \frac{1}{2}\frac{\left(\operatorname{grad}\varphi(z^{(k)}), p^{(k)}\right)}{(Qp^{(k)}, p^{(k)})}.$$

Step 1.4. Compute $a_{\max}$, the maximum possible stepsize along this direction $p^{(k)}$ without going outside of Y.

Step 1.5. If $a_k \le a_{\max}$, then set $z^{(k+1)} = z^{(k)} + a_k p^{(k)}$, $k = k+1$ and go to Step 1.1; otherwise $z^{(k+1)} = z^{(k)} + a_{\max}p^{(k)}$, and go to Step 1.2.

Step 2. New active constraints have arisen. Find one of these and include it into the set $I(z)$ of active constraints. In accordance with this, change the matrix of active constraints; set $m = m + 1$.

Step 3. Compute the projection P_I onto the subspace $\mathbf{R}^{n-m}$ defined by $F_I z = 0$, by the formula

$$P_I = E - F_I^*(F_I F_I^*)^{-1}F_I.$$

Step 4. Repeat Step 1, taking $z^{(k)}$ as initial point and the projection $P_I \operatorname{grad}\varphi(z^{(k)})$ instead of $\varphi(z^{(k)})$. The minimum point on the $(n-m)$-dimensional linear manifold is found in $n - m$ steps by the method of conjugate gradients. Therefore we come out of Step 4 with $k = n - m$.

If the minimum on the manifold has been found and $m = 0$, go to Step 6. If the minimum has been found and $m \ne 0$, go to Step 5. If the minimum has not been found (i.e. $a_k > a_{\max}$), go to Step 2.

Step 5. We arrive at this step only if we have found an exact minimum point on the corresponding manifold. In the sequel, displacement in this manifold or in narrower ones, obtained by adding new constraints, does not make it possible to diminish the value of $\varphi(z)$. It is necessary to relinquish one of the constraints such that when moving in the direction which has now become possible, the functional $\varphi(z)$ decreases while the point does not leave the admissible set Y. To this end we proceed as follows.

Step 5.1. Compute a set of m shadow parameters by the formula

$$u^0 = (F_I F_I^*)^{-1}F_I \operatorname{grad}\varphi.$$

Step 5.2. If all $u_i^0 \ge 0$ $(i = 1, \ldots, m)$, a solution of the problem has been found (not one constraint can be taken out of the set of active constraints). Go to Step 6.

Step 5.3. If $u_i^0 < 0$ for some i, then the ith active constraint can be excluded from $I(z)$. Return to Step 3, putting $m = m - 1$.

Step 6. End.

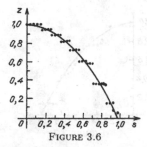

FIGURE 3.6

As is clear from the algorithm, at each transition to a new subspace we have to compute the projection operator P_I. According to the algorithm, to this end we have to invert the matrix $F_I F_I^*$. The same problem occurs when computing the shadow parameters u^0. Clearly, this is possible only if the rows of F_I are linearly independent. If the rows of the matrix of active constraints are independent, the algorithm given above solves the problem of minimizing the quadratic function $\varphi(z)$ on the set (3.12) in finitely many steps [145], [146].

We check whether there are independent constraints defining the sets $M\downarrow_C$, $\check{M}$, $\check{M}\downarrow$. First we consider the polyhedra $\check{M}$ and $\check{M}\downarrow$. It is easy to convince oneself that some constraints defining $\check{M}$ can be taken out. Clearly, $z_{i-1} - 2z_i + z_{i+1} \leq 0$ $(i = 2, \ldots, n - 1)$ and the conditions $z_1 \geq 0$, $z_n \geq 0$ imply $z_i \geq 0$ $(i = 2, \ldots, n - 1)$. Thus, $\check{M}$ can also be given by the n conditions

$$\check{M} = \left\{ z : \begin{array}{c} z_1 \geq 0 \\ z_{i-1} - 2z_i + z_{i+1} \leq 0 \quad i = 2, \ldots, n-1 \\ z_n \geq 0 \end{array} \right\}. \tag{3.14}$$

Similarly,

$$\check{M}\downarrow = \left\{ z : \begin{array}{c} z_1 \geq z_2 \\ z_{i-1} - 2z_i + z_{i+1} \leq 0 \quad i = 2, \ldots, n-1 \\ z_n \geq 0 \end{array} \right\}. \tag{3.15}$$

In both cases all remaining conditions are linearly independent, therefore the rows of all matrices F_I that may appear when realizing the method of projection of conjugate gradients will be linearly independent.

If the solution is looked for in $M\downarrow_C$, we are left with the $n + 1$ constraints

$$M\downarrow_C = \left\{ z : \begin{array}{c} z_1 \leq C \\ z_i \geq z_{i+1} \quad i = 1, \ldots, n-1 \\ z_n \geq 0 \end{array} \right\}. \tag{3.16}$$

However, it is easy to prove that for any choice of k constraints, $k \leq n$, the rows of the matrix of active constraints will be linearly independent. All $n + 1$ constraints cannot be simultaneously active, since this would lead to $C = 0$.

Thus, having made the indicated changes in the constraints defining the sets $M\downarrow_C$, $\check{M}$, $\check{M}\downarrow$, the problem of constructing a sequence minimizing $\varphi(z)$ on one of these sets

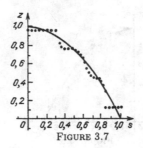

FIGURE 3.7

can be solved by the method of projection of conjugate gradients. In Chapter 4, §4 we have given a description of the programs implementing the algorithm described.

We illustrate the potential of the method of projection of conjugate gradients by solving some model problems.

In Figure 3.6 we have drawn the exact and an approximate solution of the model problem (3.6) with a priori information regarding the monotone decrease of the unknown function and the upper boundedness of this function by $C = 1$. The exact solution is $\overline{z}(s) = 1 - s^2$. We have solved the problem with exact initial information, $\delta = 0$. The approximate solution obtained corresponds to the discrepancy value $\Phi(z) = 9.28 \cdot 10^{-12}$. As initial approximation we have used the function $z^{(0)}(s) = 0.5$.

Problem (3.6) with exact solution $\overline{z}(s) = 1 - s^2$ but with righthand side perturbed by a random error such that $\delta^2 = 6.63 \cdot 10^{-5}$ has the approximate solution depicted in Figure 3.7. This solution corresponds to the following value of the discrepancy functional: $\Phi(z) = 6.62 \cdot 10^{-5}$, i.e. approximately 3% of the maximum of the righthand side. 23 great cycles were needed to obtain the solution.

3. Application of the method of projection of conjugate gradients, with projection into the set of vectors with nonnegative components, to the solution of ill-posed problems on sets of special structure

We return to the problem of finding an approximate solution of the ill-posed problem (3.1) on the sets $Z\!\downarrow$, $\check{Z}$, $\check{Z}\!\downarrow$. In §2 we have proposed an algorithm for constructing in finitely many steps an approximate solution of the problem (3.1) with a linear operator A. However, when implementing this algorithm we have to invert the matrix $F_I F_I^*$, which leads to a considerable expenditure of computer time if there are sufficiently many active constraints. In this section we propose an algorithm based on the method of projection of conjugate gradients, with projection into the set of vectors with nonnegative components. In this case the projector can be computed in an elementary way.

We will first consider the problem of constructing a minimizing sequence for the quadratic functional $\varphi(z)$, $z \in \mathbf{R}^n$, on one of the sets $\check{M}$ or $\check{M}\!\downarrow$. We will prove a number of lemmas first.

Let Y be one of $\check{M}$ or $\check{M}\!\downarrow$, and let $T^{(j)}$ $(j = 0, \dots, n)$ be the corresponding vertices of the convex bounded polyhedra $\check{M}'_C$, $\check{M}\!\downarrow$ (see Chapter 2, §3).

LEMMA 3.2. *Let $z \in Y$. Then there is a unique representation*

$$z = \sum_{j=1}^{n} a_j T^{(j)},$$

and $a_j \geq 0$ $(j = 1, \ldots, n)$.

PROOF. The truth of the lemma follows from the linear independence of the vectors $T^{(j)}$ $(j = 1, \ldots, n)$. That the a_j are nonnegative follows from their explicit representation (see Chapter 2, §3). $\square$

Consider the operator $T: \mathbf{R}^n \to \mathbf{R}^n$ defined by

$$Tx = \sum_{j=1}^{n} x_j T^{(j)}, \qquad x \in \mathbf{R}^n.$$

Since the $T^{(j)}$ $(j = 1, \ldots, n)$ form a basis in $\mathbf{R}^n$, the inverse operator T^{-1} exists, and is defined on all of $\mathbf{R}^n$. Consider the set of vectors $\Pi^+ \subset \mathbf{R}^n$ having nonnegative coordinates: $x \in \Pi^+$ if $x_j \geq 0$ $(j = 1, \ldots, n)$. Clearly, $T\Pi^+ = Y$ and $T^{-1}Y = \Pi^+$. Consider now the function $f(x) = \varphi(Tx)$, which is defined on Π^+. Instead of solving the problem of minimizing $\varphi(z)$ on Y, we may solve the problem of minimizing $f(x)$ on Π^+, since T^{-1} is a bijection between Y and Π^+.

Since T is linear, $f(x)$ is a quadratic function, as is $\varphi(z)$. Thus, the problem of finding an approximate solution of the ill-posed problem (3.1) on $\breve{Z}$ or $\breve{Z}\!\downarrow$ can be reduced to the construction of a sequence minimizing the quadratic function $f(x)$ on the set of vectors with nonnegative components.

We can similarly solve the problem of constructing an approximate solution of (3.1) on the set of monotone functions, when the requirement that the functions in $Z\!\downarrow_C$ are bounded above by a constant C is dropped.

Consider

$$M\!\downarrow = \left\{ z : z \in \mathbf{R}^n, \begin{array}{ll} z_{i+1} - z_i \leq 0, & i = 1, \ldots, n-1 \\ z_i \geq 0, & i = 1, \ldots, n \end{array} \right\}.$$

It is easy to get convinced of the fact that in this case T is a bijection from Π^+ onto $M\!\downarrow$ ($T^{(j)}$, $j = 1, \ldots, n$, the vertices of the convex bounded polyhedron $M\!\downarrow_C$, are defined by (3.7)).[4] It remains to note that the method of projection of conjugate gradients can be relatively easily implemented in Π^+ [146], since the projector into Π^+ can be trivially constructed. Recall that we need to find an element $x_\delta \in \Pi^+$ such that $f(x) \leq \delta^2$. The approximate solution z_δ of (3.1) can then be found by $z_\delta = Tx_\delta$.

The method described in this section has been proposed in [65], [67]; in Chapter 4, §5 we will describe a program implementing it.

We will illustrate the potential of the method of projection of conjugate gradients with projection into the set of vectors with nonnegative components.

---

[4]Note that if the exact solution $\bar{z}$ is monotone and bounded, then to prove convergence of the approximations to the exact solution it is essential to have information regarding the uniform boundedness of the functions in $Z\!\downarrow_C$. Here we lack this information.

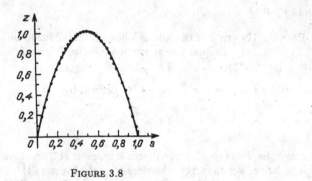

FIGURE 3.8

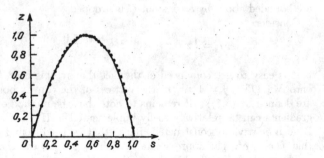

FIGURE 3.9

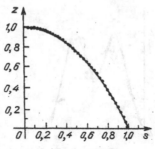

FIGURE 3.10

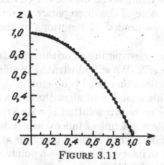

FIGURE 3.11

We take (3.6) as model problem. The exact solution is $\overline{z}(s) = 4s(1 - s)$. We have used a priori information regarding the convexity of this solution. We take the exactness of specifying the righthand side equal to zero. The approximate solution obtained is depicted in Figure 3.8 using dots. The discrepancy value is $\Phi(z) = 1.05 \cdot 10^{-14}$, which corresponds to a relative error of approximately $5 \cdot 10^{-5}\%$ (of the maximum of the righthand side). To find this solution we needed 42 iterations in total. We have taken $z^{(0)} = 0$ as initial approximation.

In Figure 3.9 we have depicted an approximate solution of (3.6) with perturbed righthand side. We have chosen the perturbation such that $\delta^2 = 7.40 \cdot 10^{-5}$, which is 3% of the maximum of the righthand side. The approximate solution has been obtained after 2 iterations by the method of projection of conjugate gradients, and corresponds to the discrepancy value $\Phi(z) = 7.29 \cdot 10^{-5}$.

The approximate solution of (3.6) on the set of monotonically decreasing, upper convex functions, with exact solution $\overline{z}(s) = 1 - s^2$ (the error of specifying the righthand side is taken to be zero), is depicted in Figure 3.10. This solution corresponds to a 'sharp' minimum (up to rounding errors in the program) of the discrepancy functional $\Phi(z)$. To find this solution we needed 224 iterations by the method of projection of conjugate gradients. The value of the discrepancy functional on the approximate solution is $\Phi(z) = 1.69 \cdot 10^{-12}$. This corresponds to an error of about 0.0005% of the maximum of the righthand side. The initial approximation was $z^{(0)} = 0$.

In Figure 3.11 we have depicted the approximate solution obtained by minimizing

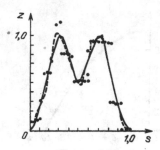

FIGURE 3.12

the corresponding discrepancy functional on the set of monotonically nonincreasing nonnegative functions. The value of the discrepancy on this solution is $\Phi(z) = 3.37 \cdot 10^{-11}$. To find this solution we needed 42 iterations by the method of projection of conjugate gradients.

Thus, we have succeeded in constructing algorithms for the approximate solution of the problems posed in Chapter 2. We will illustrate the efficiency of these algorithms in Chapter 4, using computations on model problems.

Note that now we have this complex of algorithms, we can solve also more complicated problems, when the unknown solution is only known to have, e.g., a given amount of extrema and a given amount of inflection points. Denote the latter by s_i $(i = 1, \ldots, m)$. If their location is known, the algorithms above can be adapted without difficulty to this case (in an interval between two points s_i the unknown function is monotone or convex, respectively). Under the assumptions made in the program, ill-posed problems of the type (3.1) on the set of piecewise monotone (piecewise convex) functions can be solved.

The problem becomes essentially more complicated when the location of the s_i is not known.

The great efficiency of the algorithms based on methods of conjugate gradient type and enabling the solution of problems with fixed s_i in several seconds makes it possible to solve the problem of choosing the s_i $(i = 1, \ldots, m)$ for relatively small m by a simple enumeration.

Interesting algorithms making it possible to conduct a successful enumeration of the extremum and inflection points have been given in [155].

In Figure 3.12 we have given the results of a computation of the model problem (3.6) on the set of piecewise monotone (piecewise convex) functions. As initial solution for the iteration process we have taken the function identically equal to zero.

CHAPTER 4

Algorithms and programs for solving linear ill-posed problems

In this chapter we give a description of the programs implementing the algorithms considered in this book. This description is accompanied by examples of computations of model problems. These examples are meant to serve two purposes. First, they have to illustrate the description of the program. Secondly, they have to convince a reader who wants to use our programs that these programs have been correctly inputted by him/her on his/her computer. In this case the model computations may serve as a test case for checking this.

It makes sense to first give some general remarks on the structure of the programs and on the aims calling for this structure. The problems considered in Chapter 1 and Chapter 2 lead, in the majority of cases, to the solution of a system of linear equations which are in some way or other related. It is at this stage that most complicated and laborious problems arise. In accordance with this, the kernel of the software package consists of subroutines implementing various algorithms for solving, in general, badly conditioned systems of linear equations. Nevertheless, we have to keep in mind that in the majority of cases the problem of solving systems of equations is of secondary nature: in the processing of experimental information of various kinds and in the solution of the synthesis or interpretation problem, problems of a different level arise. Most often, problems of this kind lead to a first-kind Fredholm integral equation:

$$\int_a^b K(x,s)z(s)\,ds = u(x), \qquad c \le x \le d. \tag{4.1}$$

To ensure the applicability of the software in various concrete applications, we have thought it appropriate to give programs that are tuned especially to solving the integral equations (4.1). However, the programs implementing the solution of badly conditioned systems of linear equations (including such when a priori constraints are present) are the ones most interesting. In relation with this, main attention has been given to precisely these programs, irrespective of the fact that they are, as a rule, not the main programs. We have assumed that the interested reader has no difficulty in extracting from the given units any configuration of programs he/she may need, and can supplement this configuration with any other programs that may be required.

All programs are written in FORTRAN. Moreover, the possibilities of this language have been deliberately restricted. In writing the programs, additional restrictions have been imposed upon them, making it possible to include the programs in practically any standard FORTRAN library. In the majority of cases these restrictions are not too heavy, and to satisfy them we did not have to complicate the programs in an essential way. All programs have been endowed with a sufficient amount of comments.

The main programs contain among their parameters an output-code, characterizing the result of functioning of the program. To enable the convenient usage of the main programs we have tried to keep the amount of parameters a minimum, combining all work arrays into a single one.

The text of the main programs has been imbedded into corresponding applications. Moreover, certain small subroutines used in various algorithms have been placed in the separate Appendix 9. A list of these, together with a description, can be found below.

0.1. Some general programs. A sufficiently detailed description of the programs listed below can be found in the comments in them:

PTICR0 one-dimensional minimization of a smoothing functional;
PTICR∅ computation of the matrix of an operator;
PTICR1 array moving;
PTICI2 filling an array by an integer;
PTICR2 filing an array by a REAL value;
PTICR3 multiplication of a matrix by a vector;
PTICR4 computation of the gradient of the norm $\varphi(z) = \|Az - u_\delta\|^2_{\mathbf{R}^m}$;
PTICR5 computation of the discrepancy $\varphi(z)$;
PTICR6 computation of the inner product of two vectors;
PTICR7 computation of a weighted inner product;
PTICR8 computation of the gradient of a stabilizer;
PTICR9 computation of the value of the stabilizing functional $\|z\|^2_{W_2^1}$.

In this Chapter we will also give the results of test computations. For the majority of methods, we have taken (4.1) as model equation, with kernel

$$K(x,s) = \frac{1}{1 + 100(x-s)^2} \qquad (4.2)$$

and $a = 0$, $b = 1$, $c = -2$, $d = 2$.

We have chosen the values of the righthand side $u(x)$ of (4.1) on the grid $\{x_i\}_{i=1}^m$ in $[c, d]$ in accordance with the following rule.

As vector u at the righthand side we have used the vector obtained by multiplication of the $(m \times n)$-dimensional matrix A, approximating the operator in (4.1), by the column vector $\overline{z}$ of values of the exact solution on the grid $\{s_i\}_{i=1}^n$ in the interval $[a, b]$:

$$u_i = \sum_{j=1}^{n} A_{ij}\overline{z}(s_j). \qquad (4.3)$$

This way of choosing the righthand side guarantees that the minimum of the discrepancy functional $\Phi(z) = \|Az - u_\delta\|^2$ on this set of vectors will be zero. This property of the solution is essential when using properties of iteration algorithms for solving ill-posed problems.

We will dwell on one essential factor which is characteristic of iteration algorithms for solving ill-posed problems (this concerns in the first place algorithms for solving ill-posed problems on compact sets). A common property of the majority of iteration algorithms is the rapid decline of the speed of minimization when approaching a minimum point of the functional. Therefore an important characteristic of iteration algorithms is the actual minimal level of values of the discrepancy functional up to which the minimization process runs in real time. This parameter makes it possible to estimate beforehand the error of specifying the initial information for which it makes sense the apply the given method. Or, conversely, to choose on the basis of the error of specifying the initial information an algorithm that is most suitable for solving the given problem. So, if in model problems we can minimize in real time the discrepancy functional with level $\sim 1\%$ (in relation to the norm of the righthand side), then it is clear that when using this algorithm we can, in general, successfully solve problems in which the error of specifying the initial information is $\sim 0.1\%$. In fact, such errors of specifying the initial data are characteristic in certain high-frequency astrophysical investigations.

For this reason we study the level up to which we can minimize the discrepancy and use the righthand side of (4.1) computed in accordance with (4.3).

Note that in the programs described below the solution of Fredholm integral equations of the first kind leads to a problem for (in general, badly conditioned) systems of linear equations, by approximating the integral in (4.1) by the trapezium formula on the uniform grid $\{s_j\}_{j=1}^n$ in $[a, b]$ for each value of the argument x in a grid $\{x\}_{i=1}^m$ which is also uniform. The matrix A of the linear operator approximating the integral operator in (4.1) can be chosen to be

$$A_{ij} = \begin{cases} h_s K(x_i, s_j), & j = 2, \ldots, n-1, \\ \frac{h_s}{2} K(x_i, s_j), & j = 1, n, \end{cases} \tag{4.4}$$

where $h_s = (b-a)/(n-1)$ is the step of the uniform grid $\{s_j\}_{j=1}^n$ in $[a, b]$, and $s_1 = a$, $s_n = b$.

If it is necessary to use more accurate approximation formulas, it suffices to change only the program PTICR0, which realizes the transition to a finite-difference problem and forms the entries of A.

In the solution of test problems the parameters are chosen such that the error of approximating (4.1) by the system of linear equations

$$\psi(z) = \left\{ h_x \sum_{i=1}^m \left[\int_a^b K(x_i, s) z(s) \, ds - \sum_{j=1}^n A_{ij} z(s_j) \right]^2 \right\}^{1/2},$$

where $h_x = (d-c)/(m-1)$ is the step of the uniform grid with respect to x in $[c, d]$, $x_1 = c$, $x_m = d$, is $\sim 10^{-4}$ on the exact solution $\psi(\bar{z})$. This value of the approximation error corresponds to the relative error $\psi(\bar{z})/\|\bar{u}\|_{L_2} = 0.001$.

In general, when solving (4.1) we always have to keep track of the fact that the error of approximating the integral in (4.1) is substantially smaller the the error δ of specifying the righthand side. For this it is necessary either to choose sufficiently dense grids, increasing the dimension of the problem and bringing about a substantial increase of computer time spenditure, or to use more exact quadrature formulas (the program PTICRØ). In the opposite case the approximation error can be studied by considering the problem with approximately given operator.

1. Description of the program for solving ill-posed problems by the regularization method

In this section we will describe the programs PTIMR and PTIZR, which implement the regularization method for solving the Fredholm integral equation (4.1) as described in Chapter 1, §2, §5, §6. Both algorithms considered use a choice of regularization parameter in accordance with the generalized discrepancy principle (Chapter 1, §2), i.e. the regularization parameter is determined by

$$\rho(\alpha) = \|A_h z^\alpha - u_\delta\|^2 - \left(\delta + h\|z^\alpha\|\right)^2 - \mu_\eta^2(u_\delta, A_h) = 0.$$

As stabilizing functional we use a first-order stabilizer, i.e. $Z = \mathcal{W}_2^1[a, b]$ and

$$\|z\|^2 = \int_a^b \left(|z(s)|^2 + |z'(s)|^2\right)\, ds.$$

This guarantees convergence of the approximate solutions to the exact solution in the metric of $\mathcal{W}_2^1[a, b]$, and hence uniform convergence.

To find the extremals of the smoothing functional $M^\alpha[z] = \|A_h z - u_\delta\|^2 + \alpha\|z\|^2$ for a fixed value of the regularization parameter α, the program PTIMR uses the algorithm for repeated solution of the system of equations approximating the Euler equations for the functional $M^\alpha[z]$,

$$A^* A z + \alpha C z = A^* u_\delta$$

(here, A is the matrix of the operator approximating the integral operator in (4.1), (z, Cz) approximates the norm $\|z\|_{\mathcal{W}_2^1}^2$, see Chapter 1, §6), which is described in Chapter 1, §6 and which is based on reducing this system to tridiagonal form.

The program PTIZR uses the conjugate gradient method for minimizing $M^\alpha[z]$. This program is, in general, slower than PTIMR.

Note that since $\|A\|_{\mathcal{W}_2^1 \to L_2} \leq \|A\|_{L_2 \to L_2}$, we can use a bound $h \geq \|A - A_h\|_{L_2 \to L_2}$ such that $h \to 0$ as measure for the error of specifying A.

To find the roots of the generalized discrepancy $\rho(\alpha)$, both programs use a modification of the chord method. In fact, note that if $h = 0$ (the operator is specified exactly), then $\sigma(\mu) = \rho(1/\mu)$ is a monotonically decreasing and convex function. In Figure 4.1 we have drawn the dependence of the generalized discrepancy $\sigma(\mu) = \rho(1/\mu)$ on a variable that is inversely proportional to the regularization parameter, for equation (4.1)

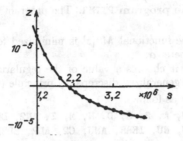

FIGURE 4.1

with kernel (4.2) and

$$\overline{z}(s) = \frac{\exp\left\{-\frac{(s-0.3)^2}{0.03}\right\} + \exp\left\{-\frac{(s-0.7)^2}{0.03}\right\}}{0.9550408} - 0.0521309113,$$

$$h^2 = 2.24 \cdot 10^{-7}, \qquad \delta^2 = 6.41 \cdot 10^{-6}.$$

If the initial approximation of the regularization parameter α is chosen sufficiently large, then $\rho(\alpha)$ will be positive.

Let $\alpha_0 = 1/\mu$, $\alpha_1 = 1/\mu_1$ be such that $\rho(\alpha_0)$, $\rho(\alpha_1)$ are positive, and suppose $\alpha_0 > \alpha_1$, i.e. $\rho(\alpha_0) > \rho(\alpha_1)$. Then the sequence given by the chord method, i.e. given by

$$\mu_n = \mu_0 - \frac{\mu_0 - \mu_1}{\sigma(\mu_0) - \sigma(\mu_1)}\sigma(\mu_0), \qquad \alpha_n = \frac{1}{\mu_n}, \qquad \mu_0 = \mu_1, \quad \mu_1 = \mu_n, \quad (4.5)$$

will be monotone: $\mu_{n+1} > \mu_n$, and $\sigma(\mu_n) \geq 0$ for all n. In this case the convergence of the chord method is guaranteed.

If, however, $h \neq 0$ and the operator is given with an error, then the convexity of the generalized discrepancy $\sigma(\mu) = \rho(1/\mu)$ can be violated and monotonicity and convergence of the sequence provided by the chord method cannot be guaranteed in that case. However, if $\sigma(\mu_n) = \rho(1/\mu_n) > 0$ for all μ_n as constructed by (4.5), then by the monotonicity of the generalized discrepancy, $\mu_{n+1} > \mu_n$. Thus, if the generalized discrepancy is positive at all μ_n, we can assert that the chord method converges. If, however, $\sigma(\mu_{n_0}) < 0$ for some n_0, then the subsequent application of the chord method according to (4.5) is not justified. Hence in this case, to solve the equation $\sigma(\mu) = 0$ for $n > n_0$ we can apply the following modification of the chord method:

$$\mu_n = \mu_0 - \frac{\mu_0 - \mu_1}{\sigma(\mu_0) - \sigma(\mu_1)}, \qquad \alpha_n = \frac{1}{\mu_n},$$

if $\sigma(\mu_0)\sigma(\mu_n) < 0$, then $\mu_1 = \mu_n$;

if $\sigma(\mu_1)\sigma(\mu_n) < 0$, then $\mu_0 = \mu_n$.

$$(4.6)$$

The fulfillment of one of these inequalities is guaranteed.

1.1. Description of the program PTIMR.

The program PTIMR may be used in two modes:

1) in the first mode the functional $M^\alpha[z]$ is minimized for a fixed value of the regularization parameter α;
2) in the second mode it chooses a value of the regularization parameter α in accordance with the generalized discrepancy principle (Chapter 1, §2).

The program is called for as follows:

CALL PTIMR(AK, U∅, A, B, C, D, N, M, Z, AN2, DL, H,
* C1, IMAX, ALFA, U, NU, IERR, AMU, C2, ALP, EPSF)

Here:

AK(X,S) is the function-subroutine for computing the kernel $K(x,s)$ of (4.1);

U∅ is an input parameter; it is the array of values of the righthand side of (4.1) on the uniform grid $\{x_i\}_{i=1}^m$ in $[c,d]$, $x_1 = c$, $x_m = d$. The grid consists of M points.

A, B, C, D are input parameters; they are equal to the quantities a, b, c, d in (4.1), respectively.

N is an input parameter; it is the dimension of the uniform grid $\{s_j\}_{j=1}^n$ in $[a,b]$, $s_1 = a$, $s_n = b$, on which the unknown function is looked for.

M is an input parameter; it is the dimension of the grid on which the righthand side of (4.1) is given (the value of M should not be less than that of N).

Z is an output parameter; it is an array of N numbers; after the program PTIMR has finished, the array Z will contain the extremal of the functional $M^\alpha[z]$ for the value ALFA of the regularization parameter.

AN2 is an output parameter; after the program has finished, it will contain the value of the discrepancy functional $\Phi(z^\alpha) = \|A_h z^\alpha - u_\delta\|_{L_2}^2$ at the extremal for $M^\alpha[z]$ found.

DL is an input parameter; it is the value of the error δ^2 of specifying the righthand side of (4.1): $\delta^2 \geq \|u_\delta - \overline{u}\|_{L_2}^2$.

H is an input parameter; it is the value of the error h^2 of specifying the operator A of (4.1): $h^2 \geq \|A - A_h\|_{L_2}^2$.

C1 is a real input parameter; it determines the mode of functioning of the program PTIMR:

C1 $\leq$ 1.0: determine an extremal of $M^\alpha[z]$ for a fixed value ALFA of the regularization parameter;

C1 $>$ 1.0: choose the regularization parameter in accordance with the generalized discrepancy principle, i.e. from the equation $\rho(\alpha) = 0$; in this case the parameter C1 determines also the accuracy of the solution of the discrepancy equation, in fact, we look for an α such that $|\rho(\alpha)| \leq \epsilon = (c_1 - 1)\delta^2$, where c_1 is the value of C1.

IMAX is an input parameter; it is the maximum number of iterations of the chord method allowed when solving the discrepancy equation; when choosing the regularization parameter, it is also the maximum number of allowed changes of this parameter for which the generalized discrepancy is positive.

ALFA is an input and an output parameter; it is the regularization parameter;

- in the first mode ($C1 \leq 1.0$) it contains the value of the regularization parameter for which one looks for an extremal of $M^{\alpha}[z]$;
- in the second mode ($C1 > 1.0$) it serves as initial approximation for a root of the equation $\rho(\alpha) = 0$. We recommend to choose the initial approximation such that at it the value of the generalized discrepancy be positive. If $\|u_{\delta}\|^2 > \delta^2$, this condition will be satisfied for all sufficiently large values of the regularization parameter α. If the value of the generalized discrepancy at a given initial approximation of the regularization parameter is negative, we multiply this initial approximation by 2 as long as the generalized discrepancy does not become positive at it (but at most IMAX times). To construct a sequence by the chord method, we use as second point a value of the regularization parameter that is twice as large as that stored in ALFA;
- in the second mode of choosing the regularization parameter ($C1 > 1.0$), when PTIMR has finished ALFA contains the required root of the discrepancy equation.

U is a work array of length at least $N * M + 10 * N + 2 * M$.

NU is an input parameter; it is the length of the work array U.

IERR is an output parameter; it is the code with which the program has finished.

IERR $= 0$ if we have found a value for the regularization parameter that satisfies with given accuracy the generalized discrepancy principle, if this is what was required ($C1 > 1.0$) (only the version (4.5) of the chord method has been used), or we have found an extremal of $M^{\alpha}[z]$ for the fixed value ALFA of the regularization parameter ($C1 \leq 1.0$).

IERR $= 1$ if we have found a value for the regularization parameter that satisfies the generalized discrepancy principle, while in performing the chord method we have encountered negative values of the generalized discrepancy, i.e. the convexity of the functional $\sigma(\mu) = \rho(1/\mu)$ is violated (we have used version (4.6) of the chord method).

IERR $= 64$ if the length of the array U does not suffice for allocating the work arrays; the values of all output parameters are not defined.

IERR $= 65$ if, when choosing a value of the regularization parameter α such that $\rho(\alpha) > 0$, the value of α has been multiplied by 2 IMAX times, but the value of $\rho(\alpha)$ remained negative. Z contains an extremal of $M^{\alpha}[z]$ for value of the regularization parameter equal to ALFA.

IERR $= 66$ if, when solving the discrepancy equation $\rho(\alpha) = 0$, we have performed IMAX iterations of the chord method in accordance with (4.5), but the required accuracy has not been attained; when the program PTIMR has finished, Z contains an extremal of $M^{\alpha}[z]$ for the current value of ALFA.

IERR $= 67$ if, when solving the discrepancy equation $\rho(\alpha) = 0$, we have performed IMAX iterations of the chord method in accordance with (4.5) and (4.6). The convexity of $\sigma(\mu) = \rho(1/\mu)$ is violated (in the chord method we have encountered negative values of the generalized discrepancy); Z contains an extremal of $M^\alpha[z]$ for the current value of ALFA.

IERR $= 68$ if, in the process of solving the discrepancy equation, we have obtained a zero value for the regularization parameter; Z contains an extremal of $M^0[z]$.

IERR $= 69$ if after IMAX multiplications of α by 2 we have not been able to localize a minimum of $\rho_1(\alpha) = \|Az^\alpha - u\| + h\|z^\alpha\|$.

IERR $= 70$ if we have not found a minimum of $\rho_1(\alpha)$ with in the given accuracy.

The remaining four parameters are used only if C2 ≥ 1.

AMU is an output parameter; it contains the minimum of $\rho_1(\alpha)$ found by the subroutine AUMINM.

C2 is a real input parameter; it determines the functioning of AUMINM:
- if C2 ≥ 1, then AUMINM has found a minimum of $\rho_1(\alpha)$; the minimum is admissible if at the nth iteration $|\alpha_n - \alpha_{n-1}| \leq 1/C2$ or $|\rho_1(\alpha_n) - \rho_1(\alpha_{n-1})| \leq$EPSF.

ALP is an input and an output parameter; when reverting to PTIMR it gives an initial approximation α for the procedure of minimizing $\rho_1(\alpha)$ using AUMINM. After the program has finished, it contains the computed value of α on which $\rho_1(\alpha)$ attains its minimum.

EPSF is an input parameter; it gives the accuracy of computing the minimum for $\rho_1(\alpha)$: $|\rho_1(\alpha_n) - \rho_1(\alpha_{n-1})| \leq$EPSF.

When solving ill-posed problems for Fredholm equations of the first kind, it may turn out to be necessary to determine an extremal of $M^\alpha[z]$ for various values of α and one and the same righthand side of (4.1). The direct use of PTIMR to this end involves the repeated transformation of one and the same system of linear equations to tridiagonal form. To escape from these repeated transformations, there is the entrance PTIMRE to the program PTIMR. When using this route it is necessary to save and restore before entering the contents of the array U. Note that using PTIMRE is also possible when choosing the regularization parameter, e.g. in case PTIMR has finished with code IERR $= 65, 66, 67$.

The following subroutines are called upon: PTIMRC, PTIMRS, PTIMRD, PTIMRØ, PTIMR1 PTIMRP, PTIMRQ, PTIMRR, PTIMRN, PTIMRA, PTIMRK, PTICRØ, PTICR1, PTICR6, AUMINM.

We give a short description of the functioning of the program PTIMR. The lines 11–36 initialize the work array and also restore certain tests and iteration counters. Then, if this is necessary (we have not entered via PTIMRE) the successive calls of subroutines perform the following actions.

1. In the array AK the matrix A of the linear operator approximating the integral operator is stored (subroutine PTICRØ).

2. We form the diagonal C and subdiagonal B of the matrix of the stabilizer:

$$C = \begin{pmatrix} 1+\frac{1}{h^2} & -\frac{1}{h^2} & 0 & & & \\ -\frac{1}{h^2} & 1+\frac{2}{h^2} & -\frac{1}{h^2} & 0 & & \\ \cdots & \cdots & \cdots & \cdots & & \cdots \\ & 0 & & -\frac{1}{h^2} & 1+\frac{2}{h^2} & -\frac{1}{h^2} \\ & & & 0 & -\frac{1}{h^2} & 1+\frac{1}{h^2} \end{pmatrix}.$$

3. The matrix constructed is written in the form $C = S'S$, with S a bidiagonal matrix (S1 is its diagonal, S2 its supdiagonal). We use the square-root method to obtain this representation.

4. The subroutine PTIMRD transforms the matrix of the system, multiplying it on the right by the inverse of S: $A = AS^{-1}$.

5. The subroutine PTIMRØ writes A as QPR, with Q, R orthogonal matrices of dimensions $m \times m$ and $n \times n$, respectively, and P is an upper bidiagonal matrix. To construct this decomposition we use the algorithm described in Chapter 1, §6, which is based on the successive application of the operation of reflection of rows and columns to the initial matrix. The subroutine PTIMRØ writes the diagonal and subdiagonal of P into the arrays P1 and P2, respectively, which are both of length n. As already noted in Chapter 1, it is not necessary to store explicitly the orthogonal matrices Q and R, which are equal to $Q = Q_1 \ldots Q_n$ and $R = R_{n-2} \ldots R_1$, respectively, where Q_i and R_i are the matrices of the reflection operators in the rows and columns. Each of these matrices is determined by a unique vector normal to a hyperplane in $\mathbf{R}^m$, $\mathbf{R}^n$, respectively. In making up Q_i, R_i in accordance with the algorithm in Chapter 1, §6, the number of nonzero entries in these vectors is $m - i + 1$, $n - i$, respectively. This makes it possible to replace the $(m \times n)$-dimensional matrix A by the reflection vectors defining Q_i and R_i; PTIMRØ thus does not store A. Let w_{ik} $(i = 1, \ldots, m)$ denote the elements of the reflection vector defining Q_k $(k = 1, \ldots, n)$, and v_{jk} $(j = 1, \ldots, n)$ the elements of the vector defining R_k, then the replacement of A by these vectors can be written as follows:

$$A = \begin{pmatrix} w_{11} & v_{12} & \cdots & v_{n-11} & v_{n1} \\ w_{12} & w_{22} & \cdots & v_{n-12} & v_{n2} \\ \cdots & \cdots & \cdots & \cdots & \cdots \\ w_{n-11} & w_{n-12} & \cdots & w_{n-1n-1} & 0 \\ w_{n1} & w_{n2} & \cdots & w_{nn-1} & w_{nn} \\ \cdots & \cdots & \cdots & \cdots & \cdots \\ w_{m1} & w_{m2} & \cdots & w_{mn-1} & w_{mn} \end{pmatrix},$$

where the elements not written out are equal to zero.

When implementing the algorithm we have to take into account the possibility of accumulation of rounding errors, as well as the possibility that machine zeros appeared. In relation with this we have used special tools to compute inner products and normalization of vectors. In particular, we propose the program PTIMR1 for computing the norms of vectors; in it a preliminary normalization is performed.

6. The subroutine **PTIMRQ** computes $q = Q'u_\delta$ and stores it in the array **U2**. Here, u_δ is the righthand side of the integral equation (4.1). The vector **U2** is subsequently used in **PTIMRA** for computing the discrepancy. When multiplying Q from the right by a vector, this vector is subsequently reflected using the vectors stored as the columns of the matrix **AK**.

7. The subroutine **PTIMRP** makes up the tridiagonal matrix $P'P$. The arrays **C**, **B**, **A** contain, respectively, the subdiagonal, the diagonal, and the supdiagonal of this matrix.

8. Finally, **PTIMRR** stores in **U1** the vector $f = RD'u_\delta$, by multiplying the vector q, which is stored in **U2** and has been computed by **PTIMRQ**, by the bidiagonal matrix P'. The vector f serves as righthand side of the tridiagonal system of equations $(P'P + \alpha E)x = f$.

The lines 47–51 assign to **AN1** the value $\sum_{i=n+1}^{m} q_i^2$. This variable enters as a summand in the discrepancy, independent of the regularization parameter α. It is a lower estimate for the measure of compatibility of the initial system of equations and is used in the program **PTIMRA** for computing the generalized discrepancy as measure of the incompatibility.

If an error $h \neq 0$ is involved in specifying the operator, then we have to minimize $\rho_1(\alpha) = \|Az^\alpha - u\| + h\|z\|$ to find it. To this end we first multiply α by 2 to localize a minimum α_0 of $\rho_1(\alpha)$ from the right and the left, and then construct a sequence α_n converging to α_0 by the golden section method. The computation ends if $|\alpha_n - \alpha_{n-1}| \leq 1/\texttt{C2}$ or $|\rho_1(\alpha_n) - \rho_1(\alpha_{n-1})| \leq \texttt{EPSF}$.

The lines 56–58 construct the diagonal of the system $(P'P + \alpha E)x = f$ and solve this system by the sweep method (subroutine **PTIMRN**). Then the subroutine **PTIMRA** computes the discrepancy and generalized discrepancy corresponding to the given extremal **Z** obtained by the subroutine **PTIMRN**. Moreover, as has been seen in Chapter 1, it is not necessary to pass to the initial unknowns, since the discrepancy can be computed by the formula $\|Px - q\|^2$, where P is a bidiagonal matrix. As already noted, some of the summands in this norm are constant and do not depend on the regularization parameter; their sum **AN1** is computed only once.

If an error $h \neq 0$ is involved in specifying the operator, then to find the generalized discrepancy we have to compute the norm of the solution in the space $\mathcal{W}_2^1$. To this end it is also not necessary to return to the old variables, since the norm of the solution in the finite-dimensional analog of $\mathcal{W}_2^1$ coincides with the norm of the solution of the tridiagonal system $(P'P + \alpha E)x = f$ in the ordinary Euclidean space.

The discrepancy and norm of the solution computed in this way are multiplied by the steps of the grids with respect to the variables s and x in (4.1) in such a way that the norms obtained are approximations of the integral norms in $L_2[c, d]$ and $\mathcal{W}_2^1[a, b]$. The operation of multiplying α by the quotient of the steps of the grids, in the formation of the diagonal of $P'P + \alpha E$ in lines 57, 72, 88, 97, has the same meaning.

The lines 60–66 verify the necessity of halting the program **PTIMR** and the fulfillment of the condition $\rho(\alpha) > 0$. If this condition does not hold, then the regularization parameter α is multiplied by 2 until $\rho(\alpha)$ becomes positive. All in all, **IMAX** of such multiplications by 2 are allowed.

The lines 67–74 make up a second point, corresponding to doubling the value of the regularization parameter, which is necessary for starting the chord method.

The lines 75–91 implement the chord method. If the value of the generalized discrepancy remains negative, we are led to line 92, the start of the modified chord method, which is implemented in lines 92–111.

The program PTIMR stops working by performing halting tests and an inverse transition to the old variables; this is implemented in the subroutine PTIMRK, as described in Chapter 1, §6.

As test computation we propose to use the solution of equation (4.1) with kernel (4.2) and vales $a = 0$, $b = 1$, $c = -2$, $d = 2$. As exact solution we use

$$\overline{z}(s) = \frac{e^{-(s-0.3)^2/0.03} + e^{-(s-0.7)^2/0.03}}{0.9550408} - 0.052130913.$$

We take the grid of the variable s on $[a, b]$ to consist of $n = 41$ points. We will use values of the righthand side on the grid with $m = 41$ points as computed by (4.3). The accuracy of specifying the righthand side and the operator are assumed to be equal to $\delta^2 = 10^{-8}$ and $h^2 = 10^{-10}$. As initial approximation of the regularization parameter we take $\alpha = 4 \cdot 10^{-4}$. We solve the discrepancy equation $\rho(\alpha) = 0$ with accuracy 10^{-11}. (i.e. we take C1 = 1.001), and look for an α such that $|\rho(\alpha)| \leq 10^{-11}$.

The numerical values of the approximate solution and program calling PTIMR are depicted in Figure 4.2.

```
00001          IMPLICIT REAL*8(A-H,O-Z)
00002          IMPLICIT INTEGER*4(I-N)
00003          DIMENSION ZO(41),UO(41),U(5000),Z(41)
00004          EXTERNAL AK
00005          X1=0.
00006          X2=1.
00007          Y1=-2.
00008          Y2=2.
00009          N=41
00010          M=41
00011          IMAX=1000
00012          C1=1.001
00013          ALFA=0.0004
00014          HX=(X2-X1)/(N-1.)
00015          CALL PTICRO(AK,U,X1,X2,Y1,Y2,N,M)
00016          H=1.E-10
00017          DO 57 I=1,N
00018          X=X1+HX*(I-1.)
00019          ZO(I)=(DEXP(-(X-0.3)**2/0.03)+
               +DEXP(-(X-0.7)**2/0.03))/0.955040800-
               -0.0521309113
00020       57 CONTINUE
00021          CALL PTICR3(U,ZO,UO,N,M)
00022          PRINT 501,(ZO(II),II=1,N)
00023      501 FORMAT(1X,'Exact solution='/(5F11.7))
00024          DL=1.E-8
00025          PRINT 502,H,DL
00026      502 FORMAT(' .'/
               *' Errors:                 in operator         -',D16.9/
               *'                 in righthand side           -',D16.9)
00027          CALL PTIMR(AK,UO,X1,X2,Y1,Y2,N,M,Z,AN2,
               *DL,H,C1,IMAX,ALFA,U,10000,IERR)
00028          PRINT 503,(Z(II),II=1,N)
00029          PRINT 504,IERR,AN2,ALFA
00030      503 FORMAT(' .'/
               *        ' Approximate solution:'/(5F11.7))
00031      504 FORMAT(' .'/
               *        ' Return code             :',I5/
               *        ' Discrepancy             :',D16.9/
               *        ' Regularization parameter :',D16.9)
00032          STOP
00033          END
```

```
00001        FUNCTION AK(X,Y)
00002        IMPLICIT REAL*8(A-H,O-Z)
00003        AK=1./(1.+100.*(X-Y)**2)
00004        RETURN
00005        END
```

Exact solution=
```
 .0000000    .0320465    .0782459    .1415609    .2238816
 .3251418    .4425163    .5699656    .6983832    .8164924
 .9124512    .9758983   1.0000000    .9829996    .9288697
 .8468924    .7502625    .6540357    .5728488    .5188144
 .4998815    .5188144    .5728488    .6540357    .7502625
 .8468924    .9288697    .9829996   1.0000000    .9758983
 .9124512    .8164924    .6983832    .5699656    .4425163
 .3251418    .2238816    .1415609    .0782459    .0320465
 .0000000
```

Errors: in operator   -   .100000001D-09
                     in righthand side -   .999999994D-08

Approximate solution:
```
 .0173081    .0311578    .0715787    .1366338    .2238399
 .3296361    .4487824    .5745464    .6991498    .8137768
 .9084307    .9725681    .9974897    .9802096    .9256576
 .8447169    .7506126    .6567675    .5762364    .5213060
 .5017045    .5213060    .5762364    .6567675    .7506126
 .8447169    .9256576    .9802096    .9974897    .9725681
 .9084308    .8137768    .6991498    .5745464    .4487824
 .3296361    .2238399    .1366338    .0715788    .0311579
 .0173083
```

Return code   :   0
Discrepancy   :   .178224579D-07
Regularization parameter :   .244141280D-06

FIGURE 4.2

1.2. Description of the program PTIZR. Like PTIMR, the program PTIZR may also be used in two modes:

1) in the first mode the functional $M^\alpha[z]$ is minimized for a fixed value of the regularization parameter;
2) in the second mode it chooses a value of the regularization parameter α in accordance with the generalized discrepancy principle $\rho(\alpha) = 0$ (Chapter 1, §2).

The program is called for as follows:

```
CALL PTIZR(AK, U∅, A, B, C, D, N, M, Z, AN2, DL, H,
* C1, IMAX, ALFA, U, NU, IERR)
```

Here:

AK(X,S) is the function-subroutine for computing the kernel $K(x, s)$ of (4.1);

U∅ is an input parameter; it is the array of values of the righthand side of (4.1) on the uniform grid $\{x_i\}_{i=1}^m$ in $[c, d]$, $x_1 = c$, $x_m = d$. The grid consists of M points.

A, B, C, D are input parameters; they are equal to the quantities a, b, c, d in (4.1), respectively.

N is an input parameter; it is the dimension of the uniform grid $\{s_j\}_{j=1}^n$ in $[a, b]$, $s_1 = a$, $s_n = b$, on which the unknown function $z(s)$ is looked for.

M is an input parameter; it is the dimension of the grid $\{x_i\}_{i=1}^m$ on which the righthand side of (4.1) is given.

Z is an input and an output parameter. When calling PTIZR, Z will contain the initial approximation to the extremal of the functional $M^\alpha[z]$ for the value ALFA of the regularization parameter (it is an array of length N). When PTIZR is finished, Z will contain the array of values of the extremal of $M^\alpha[z]$ for the value ALFA of the regularization parameter

AN2 is an output parameter; after the program has finished, it will contain the value of the discrepancy functional $\Phi(z^\alpha) = \|A_h z^\alpha - u_\delta\|^2$ at the extremal Z.

DL is an input parameter; it is the value of the error δ^2 of specifying the righthand side of (4.1): $\delta^2 \geq \|u_\delta - \overline{u}\|_{L_2}^2$.

H is an input parameter; it is the value of the error h^2 of specifying the operator A of (4.1): $h^2 \geq \|A - A_h\|^2$.

C1 is a real input parameter; it determines the mode of functioning of the program PTIZR:

    C1 ≤ 1.0: determine an extremal of $M^\alpha[z]$ for a fixed value ALFA of the regularization parameter;

    C1 > 1.0: choose the regularization parameter in accordance with the generalized discrepancy principle, i.e. from the equation $\rho(\alpha) = 0$; in this case the parameter C1 determines also the accuracy of the solution of the discrepancy equation, in fact, we look for an α such that $|\rho(\alpha)| \leq \epsilon = (c_1 - 1)\delta^2$, where c_1 is the value of C1.

IMAX is an input parameter; it is the maximum number of iterations of the chord method allowed when solving the discrepancy equation; when choosing the

regularization parameter, it is also the maximum number of allowed multiplications by 2 of this parameter making the generalized discrepancy positive.
ALFA is an input and an output parameter;

- in the mode of choosing the regularization parameter (C1 > 1.0), when calling PTIZR it contains an initial approximation for the solution of the discrepancy equation. Here we recommend taking an initial approximation in ALFA for which the value of the generalized discrepancy at the extremal of the functional is positive. If the squared norm $\|u_\delta\|_{L_2}^2$ of the righthand side is larger than the value of DL, then this condition is satisfied for all sufficiently large values of the regularization parameter. If the value of the generalized discrepancy at the initial approximation is negative, then we multiply the regularization parameter by 2 until the generalized discrepancy becomes positive (but at most IMAX times). After the subroutine PTIZR has finished, ALFA contains a value of the regularization parameter satisfying the generalized discrepancy principle (this is true only if the subroutine ended successfully).
- in the mode C1 ≤ 1.0, when calling PTIZR the parameter ALFA contains a value of the regularization parameter for which we have to find an extremal of $M^\alpha[z]$.

U is a work array of length at least $N * M + 2 * N + M$.
NU is an input parameter; it is the length of the work array U.
IERR is an output parameter; it is the code with which the program has finished.

IERR = 0 if we have found a value for the regularization parameter that satisfies the generalized discrepancy principle, if this is what was required (C1 > 1.0) (only the version (4.5) of the chord method has been used), or we have found an extremal of $M^\alpha[z]$ for the fixed value ALFA of the regularization parameter (C1 ≤ 1.0).

IERR = 1 if we have found a value for the regularization parameter that satisfies the generalized discrepancy principle. When solving the discrepancy equation by the chord method we have used version (4.6) of the chord method (we have encountered negative values of the generalized discrepancy, i.e. the convexity of the functional $\sigma(\mu) = \rho(1/\mu)$ is violated); when PTIZR has finished, Z contains an extremal of $M^\alpha[z]$.

IERR = 64 if the length of the array U does not suffice; the values of all output parameters are not defined.

IERR = 65 if, when choosing a value of the regularization parameter α at which the generalized discrepancy has a positive value, the value of α has been multiplied by 2 IMAX times, but the value of the generalized discrepancy remained negative. Now Z contains an extremal of $M^\alpha[z]$ for the value of the regularization parameter stored in ALFA.

IERR = 66 if, when solving the discrepancy equation $\rho(\alpha) = 0$, we have performed IMAX iterations of the chord method in accordance with (4.5), but the required accuracy has not been attained; now Z contains an extremal of $M^\alpha[z]$ for the value α currently stored in ALFA.

> IERR = 67 if, when solving the discrepancy equation, we have performed
> IMAX iterations of the chord method in accordance with (4.5) and the
> modified chord method (4.6), but the required accuracy has not been
> attained. In the chord method we have encountered negative values
> of the generalized discrepancy. The convexity of $\sigma(\mu) = \rho(1/\mu)$ is
> violated; Z contains an extremal of $M^\alpha[z]$ for the current value of ALFA.
> IERR = 68 if we have obtained a zero value for the regularization param-
> eter; Z contains an extremal of $M^0[z]$.

When solving (4.1) it may turn out to be necessary to determine an extremal of
$M^\alpha[z]$ for various values of α. By using an entry PTIZRE we can avoid the repeated
formation of the matrix of the operator.

To this end it suffices to repeatedly call PTIZRE with the same actual parameters.
Here, between the call for PTIZR and that for PTIZRE we have to store the first $N*M$
elements of the array U. We can also use PTIZRE to continue calculations when choosing
the regularization parameter in accordance with the generalized discrepancy principle
if, e.g., PTIZR has finished with code IERR = 65, 66, or 67.

The following subroutines are called for: PTIZR1, PTIZRA, PTICRO, PTICRØ, PTICR1,
PTICR3, PTICR4, PTICR5, PTICR6, PTICR8.

We will briefly describe the functioning of the program PTIZR.

The lines 9–26 initialize the arrays, provide the necessary initial settings, and also
make up the matrix A of the operator (if necessary). The latter is done by the
subroutine PTICRØ.

In the lines 28–37 a regularization parameter α is chosen such that the value of the
generalized discrepancy corresponding to it is positive. To this end the regularization
parameter is successively multiplied by 2 (at most IMAX times). To find an extremal
of $M^\alpha[z]$ we use the subroutine PTIZR1, which implements the conjugate gradients
method. The subroutine PTIZRA computes the value of the generalized discrepancy.

In the lines 38–44 we choose a second initial point for the chord method. We have
implemented the chord method in lines 45–58. If the convexity of the generalized
discrepancy is violated, we may need the modified chord method, implemented in
lines 59–77.

Note that we use at each step an extremal of $M^\alpha[z]$ for the previous value of the
regularization parameter as initial approximation for minimizing $M^\alpha[z]$.

The subroutine PTIZR1 for minimizing $M^\alpha[z]$ performs N iterations of the conjugate
gradients method. To compute the gradient of $M^\alpha[z]$, the value of the discrepancy
functional and for one-dimensional minimization, we use the corresponding routines
PTICR4, PTICR5, PTICR8, PTICRO.

As test example of the use of PTIZR we have considered (4.1) with kernel (4.2) and
values $a = 0$, $b = 1$, $c = -2$, $d = 2$ with exact solution

$$\overline{z}(s) = \exp\left\{-\frac{(s-0.5)^2}{0.06}\right\}.$$

For both variables, we have taken grids of 41 points ($n = m = 41$). As righthand side
we have used the value computed by (4.3). We have taken the accuracy of specifying

the operator and righthand side equal to $h^2 = 10^{-10}$ and $\delta^2 = 10^{-8}$, respectively. We have taken $\alpha = 10^{-6}$ as initial approximation of the regularization parameter. We will solve the discrepancy equation with accuracy 10^{-11}, which corresponds to the value C1 $= 1.001$. The numerical values of the solution obtained are listed in Figure 4.3. This solution corresponds to the value $\alpha = 9.30 \cdot 10^{-7}$ and discrepancy equal to $1.52 \cdot 10^{-8}$.

```
00001          IMPLICIT REAL*8(A-H,O-Z)
00002          IMPLICIT INTEGER*4(I-N)
00003          DIMENSION U0(41),Z(41),U(5000),Z0(41)
00004          EXTERNAL AK
00005          X1=0.
00006          X2=1.
00007          Y1=-2.
00008          Y2=2.
00009          N=41
00010          M=41
00011          IMAX=1000
00012          C1=1.001
00013          ALFA=0.000001
00014          HX=(X2-X1)/(N-1.)
00015          CALL PTICR0(AK,U,X1,X2,Y1,Y2,N,M)
00016          H=1.E-10
00017          DO 57 I=1,N
00018          X=X1+HX*(I-1.)
00019          Z0(I)=(DEXP(-(X-0.5)**2/0.06))
00020       57 CONTINUE
00021          PRINT 501,(Z0(II),II=1,N)
00022      501 FORMAT(1X,'Exact solution='/(5F11.7))
00023          CALL PTICR3(U,Z0,U0,N,M)
00024          DL=1.E-8
00025          PRINT 502,H,DL
00026      502 FORMAT(' .'/
            *' Errors:                 in operator        -',D16.9/
            *'                 in righthand side          -',D16.9)
00027          DO 34 I=1,N
00028       34 Z(I)=0.
00029          CALL PTIZR(AK,U0,X1,X2,Y1,Y2,N,M,Z,
            *IC,AN2,DL,H,C1,ANGRD,IMAX,ALFA,U,
            *10000,IERR)
00030          PRINT 503,(Z(II),II=1,N)
00031          PRINT 504,IERR,AN2,ALFA
00032      503 FORMAT(' .'/
            *' Approximate solution:'/(5F11.7))
00033      504 FORMAT(' .'/
            *' Return code              :',I5/
            *' Discrepancy              :',D16.9/
            *' Regularization parameter :',D16.9)
00034          STOP
00035          END
```

```
00001        FUNCTION AK(X,Y)
00002        IMPLICIT REAL*8(A-H,O-Z)
00003        AK=1./(1.+100.*(X-Y)**2)
00004        RETURN
00005        END
```

Exact solution=
```
  .0155039    .0232740    .0342181    .0492711    .0694834
  .0959671    .1298122    .1719732    .2231302    .2835359
  .3528661    .4300946    .5134171    .6002454    .6872893
  .7707304    .8464817    .9105104    .9591895    .9896374
 1.0000000    .9896374    .9591895    .9105104    .8464817
  .7707304    .6872893    .6002454    .5134171    .4300946
  .3528661    .2835359    .2231302    .1719732    .1298122
  .0959671    .0694834    .0492711    .0342181    .0232740
  .0155039
```

Errors: in operator   -   .100000001D-09
                   in righthand side   -   .999999994D-08

Approximate solution:
```
  .0211643    .0240784    .0328794    .0475972    .0684072
  .0957261    .1302309    .1728098    .2243091    .2850577
  .3545593    .4316402    .5146232    .6011371    .6878841
  .7708297    .8457517    .9088284    .9568809    .9871938
  .9975960    .9871938    .9568809    .9088284    .8457517
  .7708297    .6878841    .6011371    .5146232    .4316402
  .3545593    .2850577    .2243092    .1728098    .1302309
  .0957261    .0684072    .0475972    .0328794    .0240785
  .0211644
```

Return code   :   0
Discrepancy   :   .151755068D-07
Regularization parameter :   .929922913D-06

FIGURE 4.3

2. Description of the program for solving integral equations with a priori constraints by the regularization method

In this section we will describe the program PTIPR, which is intended for solving the Fredholm integral equation (4.1) in case we a priori know that the solution is monotone or positive. The basics of the method have been described in Chapter 1 §2, §5, §6. To choose the regularization parameter we use the generalized discrepancy principle. We take the stabilizing functional to be a stabilizer of order two, i.e. we use the Hilbert space Z with norms

$$\|z\|_1^2 = \int_a^b \left(|z(s)|^2 + |z'(s)|^2 \right) ds, \qquad \|z\|_2^2 = z(b) + \int_a^b |z'(s)|^2 ds.$$

This guarantees the uniform convergence of the approximate solutions to the exact solution. For finding an extremal of the smoothing functional $M^\alpha[z] = \|A_h z - u_\delta\|^2 + \alpha\|z\|^2$ for a fixed value of the regularization parameter α, we have implemented in PTIPR the method of projection of conjugate gradients, which is described in Chapter III, §3.

To choose the regularization parameter we have implemented the modification of the chord method described in the previous section.

2.1. Description of the program PTIPR. The program PTIPR may be used in two modes:

1) in the first mode the functional $M^\alpha[z]$ is minimized for a fixed value of the regularization parameter α;
2) in the second mode it chooses a value of the regularization parameter α in accordance with the generalized discrepancy principle $\rho(\alpha) = 0$ (Chapter 1 §2).

The program is called for as follows:

```
CALL PTIPR(AK, U0, A, B, C, D, N, M, Z, IC, AN2, DL,
* H, C1, ANGRD, IMAX, ALFA, U, NU, IERR)
```

Here:

AK(X,S) is the function-subroutine for computing the kernel $K(x,s)$ of (4.1);

U0 is an input parameter; it is the array of values of the righthand side of (4.1) on the uniform grid $\{x_i\}_{i=1}^m$ in $[c,d]$, $x_1 = c$, $x_m = d$. The grid consists of points.

A, B, C, D are input parameters; they are equal to the quantities a, b, c, d in (4.1) respectively.

N is an input parameter; it is the dimension of the uniform grid $\{s_j\}_{j=1}^n$ in $[a, b]$ $s_1 = a$, $s_n = b$, on which the unknown function is looked for.

M is an input parameter; it is the dimension of the grid $\{x_i\}_{i=1}^m$ on which the righthand side of (4.1) is given.

Z is an input and an output parameter. When calling PTIPR, the array Z contains an initial approximation of the extremal of $M^\alpha[z]$ for the value of the regularization parameter stored in ALFA (it is an array of length N). After the program PTIPR has finished, the array Z will contain the extremal of $M^\alpha[z]$ for the value of the regularization parameter stored in ALFA.

IC is an input parameter; it determines the set of a priori constraints on the solution looked for.

    IC $= 0$ if the solution of (4.1) is looked for in the set of nonnegative functions $z(s) \geq 0$. We use $\|z\|_1$ as norm in Z.

    IC $= 1$ if the solution of (4.1) is looked for in the set of nonnegative monotonically nonincreasing functions. We use $\|z\|_2$ as norm in Z.

AN2 is an output parameter; after the program has finished, it will contain the value of the discrepancy functional $\Phi(z^\alpha) = \|A_h z^\alpha - u_\delta\|_{L_2}^2$ at the extremal for $M^\alpha[z]$ found.

DL is an input parameter; it is the value of the error δ^2 of specifying the righthand side of (4.1): $\delta^2 \geq \|u_\delta - \overline{u}\|_{L_2}^2$.

H is an input parameter; it is the value of the error h^2 of specifying the operator A of (4.1): $h^2 \geq \|A - A_h\|_{L_2}^2$.

C1 is a real input parameter; it determines the mode of functioning of the program PTIPR:

    C1 ≤ 1.0: determine an extremal of $M^\alpha[z]$ for a fixed value ALFA of the regularization parameter;

    C1 > 1.0: choose the regularization parameter in accordance with the generalized discrepancy principle, i.e. from the equation $\rho(\alpha) = 0$; in this case the parameter C1 determines also the accuracy of the solution of the discrepancy equation, in fact, we look for an α such that $|\rho(\alpha)| \leq \epsilon = (c_1 - 1)\delta^2$, where c_1 is the value of C1.

ANGRD is a real input parameter; it characterizes the accuracy of the solution of the problem of minimizing the smoothing functional for a fixed value of the parameter. We proceed with the minimization procedure until $\| \text{grad} \, M^\alpha[z]\|$ becomes less than the value stored in ANGRD.

IMAX is an input parameter; it is the maximum number of iterations of the chord method allowed when solving the discrepancy equation; when choosing the regularization parameter, it is also the maximum number of allowed multiplications by 2 of this parameter for which the generalized discrepancy is positive.

ALFA is an input and an output parameter.

    - in the mode of choosing the regularization parameter (C1 > 1.0), when calling PTIPR it contains an initial approximation for the root of the discrepancy equation. We recommend to choose the initial approximation such that the value of the generalized discrepancy at the extremal of $M^\alpha[z]$ be positive. If $\|u_\delta\|_{L_2}^2$ exceeds DL, this condition will be satisfied for all sufficiently large values of the regularization parameter. If the value of the generalized discrepancy at the given initial approximation is negative, we multiply this initial approximation by 2 until the generalized discrepancy becomes positive (but at most IMAX times). After finishing of PTIPR the parameter ALFA contains a value of the regularization parameter satisfying the generalized discrepancy principle (if the program is successful).

- if C1 $\leq$ 1.0, when the program is called for the parameter ALFA contains the value of the regularization parameter for which we have to determine an extremal of $M^\alpha[z]$.

U is a work array of length at least $N * M + 3 * N + 2 * M$.

NU is an input parameter; it is the length of the work array U.

IERR is an output parameter; it is the code with which the program has finished.

IERR = 0 if we have found a value for the regularization parameter that satisfies the generalized discrepancy principle, if this is what was required (only the version (4.5) of the chord method has been used), or we have found an extremal of $M^\alpha[z]$ for the fixed value ALFA of the regularization parameter (C1 $\leq$ 1.0).

IERR = 1 if we have found a value for the regularization parameter that satisfies the generalized discrepancy principle $\rho(\alpha) = 0$ with given accuracy using the modified chord method (4.6), while we have encountered negative values of the generalized discrepancy, i.e. the convexity of the function $\sigma(\mu) = \rho(1/\mu)$ is violated. When PTIPR has finished, Z contains the extremal of $M^\alpha[z]$.

IERR = 64 if the length of the array U does not suffice; the values of all output parameters are not defined.

IERR = 65 if, when choosing a value of the regularization parameter α such that the generalized discrepancy is positive, the value of α has been multiplied by 2 IMAX times, but the value of $\rho(\alpha)$ remained negative. When PTIPR has finished, Z contains an extremal of $M^\alpha[z]$ for the value of the regularization parameter stored in ALFA.

IERR = 66 if, when solving the discrepancy equation $\rho(\alpha) = 0$, we have performed IMAX iterations of the chord method in the form (4.5), but the required accuracy has not been attained; Z contains an extremal of $M^\alpha[z]$ for the current value of ALFA.

IERR = 67 if, when solving the discrepancy equation $\rho(\alpha) = 0$, we have performed IMAX iterations of the chord method in the form (4.6), but the required accuracy has not been attained. The convexity of $\sigma(\mu) = \rho(1/\mu)$ is violated (in the chord method we have encountered negative values of the generalized discrepancy); Z contains an extremal of $M^\alpha[z]$ for the current value of ALFA.

IERR = 68 if the given or obtained value of the regularization parameter is zero; Z contains an extremal of $M^0[z]$.

IERR = 69 if we have given a value of IC different from zero or one. The values of the input parameters are not determined.

When solving equation (4.1), it may turn out to be necessary to determine an extremal of $M^\alpha[z]$ for various values of α. To avoid the repeated formation of the matrix of the operator and of its transformations, there is the entry PTIPRE. To this end, when repeatedly calling for the program it suffices to call the entry PTIPRE with the same actual parameters. Here we have to save the first $N*M$ elements of the array U between the call for PTIPR and that for PTIPRE. We can also use PTIMRE when choosing the regularization parameter in accordance with the generalized discrepancy principle, e.g. in case PTIMR has finished with code IERR $= 65, 66, 67$.

In any case, when calling PTIPR and PTIPRE we have to store an admissible approximation in Z. For IC $= 0$ the initial approximation has to be nonnegative; for IC $= 1$, in addition it has to be monotonically nonincreasing.

The following subroutines are called for: PTISR1, PTISR2, PTISR3, PTISR4, PTICR0, PTICRØ, PTICR1, PTICR2, PTICI2, PTICR3, PTICR4, PTICR5A, PTICR6, PTICR7, PTICR8, PTICR9.

The functioning of the program PTIPR is similar to that of PTIZR, but instead of calling the program PTIZ1 for minimization by the conjugate gradients method, we call for the program PTISR1, which implements minimization by the method of projection of conjugate gradients. The required transformations of the matrix of the operator and the successive approximation are performed by the subroutines PTISR2 and PTISR3. In §7 we will describe the subroutines PTISR1, PTISR2 and PTISR3.

As test example for using the program we have considered the solution of (4.1) with kernel (4.2) and values $a = 0$, $b = 1$, $c = -2$, $d = 2$ with exact solution $\bar{z}(s) = 1 - s^2$. For both variables we have chosen grids of 41 points. The program, the calls, and numerical results are given in Figure 4.4.

```
00001        IMPLICIT REAL*8(A-H,O-Z)
00002        IMPLICIT INTEGER*4(I-N)
00003        DIMENSION U0(41),Z(41),U(10000),Z0(41)
00004        EXTERNAL AK
00005        X1=0.
00006        X2=1.
00007        Y1=-2.
00008        Y2=2.
00009        N=41
00010        M=41
00011        IMAX=1000
00012        C1=1.001
00013        IC=0
00014        ALFA=0.000001
00015        HX=(X2-X1)/(N-1.)
00016        CALL PTICRO(AK,U,X1,X2,Y1,Y2,N,M)
00017        H=1.E-10
00018        DO 57 I=1,N
00019        X=X1+HX*(I-1.)
00020        Z0(I)=(DEXP(-(X-0.5)**2/0.06))
00021     57 CONTINUE
00022        PRINT 501,(Z0(II),II=1,N)
00023    501 FORMAT(1X,'Exact solution='/(5F11.7))
00024        CALL PTICR3(U,Z0,U0,N,M)
00025        DL=1.D-8
00026        PRINT 502,H,DL
00027    502 FORMAT(' .'/
             *' Errors:                 in operator       -',D16.9/
             *'                  in righthand side        -',D16.9)
00028        DO 34 I=1,N
00029     34 Z(I)=0.5
00030        CALL PTIPR(AK,U0,X1,X2,Y1,Y2,N,M,Z,IC,
             *AN2,DL,H,C1,ANGRD,IMAX,ALFA,U,
             *10000,IERR)
00031        PRINT 503,(Z(II),II=1,N)
00032        PRINT 504,IERR,AN2,ALFA
00033    503 FORMAT(' .'/
             *' Approximate solution:'/(5F11.7))
00034    504 FORMAT(' .'/
             *' Return code             :',I5/
             *' Discrepance             :',D16.9/
             *' Regularization parameter :',D16.9)
00035        STOP
00036        END
```

```
00001          FUNCTION AK(X,Y)
00002          IMPLICIT REAL*8(A-H,O-Z)
00003          AK=1./(1.+100.*(X-Y)**2)
00004          RETURN
00005          END
```

Exact solution=
```
 .0155039    .0232740    .0342181    .0492711    .0694834
 .0959671    .1298122    .1719732    .2231302    .2835359
 .3528661    .4300946    .5134171    .6002454    .6872893
 .7707304    .8464817    .9105104    .9591895    .9896374
1.0000000    .9896374    .9591895    .9105104    .8464817
 .7707304    .6872893    .6002454    .5134171    .4300946
 .3528661    .2835359    .2231302    .1719732    .1298122
 .0959671    .0694834    .0492711    .0342181    .0232740
 .0155039
```

Errors: in operator   - .100000001D-09
                in righthand side   - .100000000D-07

Approximate solution:
```
 .0210491    .0239928    .0328555    .0476350    .0684788
 .0957898    .1302494    .1727642    .2242018    .2849067
 .3543894    .4314756    .5144849    .6010413    .6878428
 .7708505    .8458373    .9089764    .9570818    .9874308
 .9978459    .9874308    .9570818    .9089764    .8458374
 .7708505    .6878428    .6010413    .5144849    .4314756
 .3543894    .2849068    .2242018    .1727642    .1302494
 .0957897    .0684788    .0476350    .0328555    .0239929
 .0210492
```

Return code   :   67
Discrepancy   :   .124056379D-07
Regularization parameter :   .838278424D-06

FIGURE 4.4

3. Description of the program for solving integral equations of convolution type

In this section we will describe a program implementing the algorithm for solving the integral equations of convolution type by using the regularization method based on determining an extremal of the functional $M^\alpha[z] = \|A_h z - u_\delta\|^2_{L_2} + \alpha\|z\|^2_{W_2^1}$ and choosing the regularization parameter in accordance with the generalized discrepancy principle (Chapter 1, §2).

Consider the equation

$$\int_{-\infty}^{+\infty} K(x-s)z(s)\,ds = u(x). \tag{4.7}$$

Suppose we know that the local support of the kernel $K(t)$ is concentrated in an interval $[L_1, L_2]$. If the exact solution $\overline{z}(s)$ of (4.7) has local support in $[a, b]$, then the exact righthand side $\overline{u}(x)$ must have support in $[c, d]$ with $c = a + L_1$, $d = b + L_2$. We will assume that the approximate righthand side is known on the interval $[c, d]$. Under these assumptions we arrive at the following problem, which is equivalent to (4.7):

$$\int_a^b K(x-s)z(s)\,ds = u(x), \qquad x \in [c, d].$$

By the assumption on the location of the local support of $\overline{z}(s)$, we can write the latter equation as

$$\int_{c-(L_1+L_2)/2}^{d-(L_1+L_2)/2} K(x-s)z(s)\,ds = u(x), \qquad x \in [c, d], \tag{4.8}$$

since

$$\left[c - \frac{L_1 + L_2}{2}, d - \frac{L_1 + L_2}{2}\right] \supset [a, b].$$

It remains to note that, by the relation between the local supports of $\overline{z}(s)$ and $K(t)$, the kernel $K(t)$ can be assumed to be periodically extended from its local support $[L_1, L_2]$ with period $d - c = T$. Thus, we have arrived at the problem considered in Chapter 1, §7 (up to a trivial change of variables). Since $K(t)$ can be taken periodic with period T, to solve the problem we can use methods based on the fast Fourier transform, as described in Chapter 1.

Note that this algorithm for solving (4.8) with a periodic kernel, and hence (4.7) with kernel having compact support, gives a solution $z(s)$ on the interval

$$\left[c - \frac{L_1 + L_2}{2}, d - \frac{L_1 + L_2}{2}\right],$$

which contains the local support of the exact solution of the exact solution $\overline{z}(s)$ of (4.7).[1] The algorithm for solving (4.8) with a kernel $K(t)$ that is periodic with period $T = d - c$ as described in Chapter 1, §7 has been implemented in the programs PTIKR and PTIKR1. Here, to solve the discrepancy equation

$$\rho(\alpha) = \|A_h z^\alpha - u_\delta\|^2 - (\delta + h\|z^\alpha\|)^2 = 0$$

---

[1]There are, in general, various methods for extending the domain of integration to an interval of length $T = d - c > b - a$ when passing to (4.8).

we have used Newton's method. The derivative of $\rho(\alpha)$ needed in this method can be easily computed by (1.21). (For an expression of $\gamma'(\alpha)$ see Chapter 1, §7.) As already noted in §1 of this Chapter, since for $h = 0$ the function $\sigma(\mu) = \rho(1/\mu)$ is monotonically decreasing and convex, convergence of Newton's method is guaranteed (provided that the value $\sigma(\mu_0) = \rho(1/\mu_0)$, with μ_0 the initial approximation, is positive). In fact, the program proposed implements Newton's method for solving the equation $\sigma(\mu) = \rho(1/\mu) = 0$. Moreover, $\sigma'(\mu) = \rho'(1/\mu)/\mu^2$.

When solving (4.8) with an approximately given operator ($h \neq 0$), the convexity of $\sigma(\mu)$ can be violated and convergence of Newton's method cannot be guaranteed. Therefore, if in the solution process by Newton's method a value of the regularization parameter $\alpha = 1/\mu$ appears at which $\rho(\alpha) < 0$ (which may happen if $\sigma(\mu)$ is not convex), then we pass to the modified chord method (4.6).

As all other programs described above, the program PTIKR may be used in two modes:

1) in the first mode the smoothing functional $M^\alpha[z]$ is minimized for a fixed value of the regularization parameter α;
2) in the second mode PTIKR chooses a value of the regularization parameter α in accordance with the generalized discrepancy principle

$$\rho(\alpha) = \|A_h z^\alpha - u_\delta\|^2 - (\delta + h\|z^\alpha\|)^2 = 0,$$

where z^α is an extremal of the smoothing functional

$$M^\alpha[z] = \|A_h z - u_\delta\|^2_{L_2} + \alpha\|z\|^2_{W_2^1}.$$

3.1. Description of the program PTIKR. The program is called for as follows:

```
CALL PTIKR(AK, U∅, A, B, C, D, L1, L2, N,
* Z, AN2, DL, H, C1, IMAX, ALFA, U, NU, IERR)
```

Here:

AK(S) is the function-subroutine for computing the kernel $K(s)$ of (4.8);

U∅ is the array of values of the righthand side u_δ of (4.8) on the uniform grid $\{x_i\}_{i=1}^n$ in $[c, d]$. The grid consists of N points:

$$x_j = \frac{d-c}{n}\left(j - \frac{1}{2}\right)$$

(the number of points in the grid must be a power of two).

A, B are output parameters; they are equal to the boundaries of the interval

$$\left[c - \frac{L_1 + L_2}{2}, d - \frac{L_1 + L_2}{2}\right]$$

on which we look for an approximate solution of (4.8).

C, D are input parameters; they are equal to the quantities c, d in (4.8).

L1, L2 are input parameters; they are equal to the boundaries of the local support of the kernel $K(s)$ in (4.8). (They are variables of type REAL.)

N is an input parameter; it is the dimension of the uniform grids

$$\{x_j\}_{j=1}^n, \quad \{s_j\}_{j=1}^n \quad s_j = \frac{d-c}{n}\left(j - \frac{1}{2}\right) + c - \frac{L_1 + L_2}{2},$$

in the intervals $[c, d]$ and $[c - (L_1 + L_2)/2, d - (L_1 + L_2)/2]$, respectively, on which the righthand side of (4.8) is given and the unknown function is looked for. The number N must be of the form 2^k with k an integer.

Z is an output parameter; it is an array of length N. When PTIKR has finished, Z contains an extremal of $M^\alpha[z]$ for the value of the regularization parameter stored in ALFA

AN2 is an output parameter; after the program has finished, it will contain the value of the discrepancy functional $\Phi(z^\alpha) = \|A_h z^\alpha - u_\delta\|_{L_2}^2$ at the extremal of $M^\alpha[z]$ stored in Z.

DL is an input parameter; it is the value of the error δ^2 of specifying the righthand side of (4.8): $\delta^2 \geq \|u_\delta - \overline{u}\|_{L_2[c,d]}^2$.

H is an input parameter; it is the value of the error h^2 of specifying the operator A of (4.8), regarded as an operator from $\mathcal{W}_2^1[c - (L_1 + L_2)/2, d - (L_1 + L_2)/2]$ into $L_2[c, d]$: $h \geq \|A - A_h\|$.

C1 is an input parameter; it determines the mode of functioning of the program PTIPR:

C1 $\leq$ 1.0: determine an extremal of $M^\alpha[z]$ for a fixed value of the regularization parameter α;

C1 $>$ 1.0: choose the regularization parameter in accordance with the generalized discrepancy principle, i.e. from the equation $\rho(\alpha) = 0$; here, the equation has been solved with accuracy $\epsilon = (c_1 - 1)\delta^2$ (where c_1 is the value of C1), i.e. we look for an α such that $|\rho(\alpha)| \leq \epsilon$.

IMAX is an input parameter; it is the maximum number of iterations of Newton's method and the chord method allowed when solving the discrepancy equation $\rho(\alpha) = 0$.

ALFA holds the value of the regularization parameter. In the second mode, of choosing the regularization parameter (C1 $>$ 1.0), it contains an initial approximation for the root of the equation $\rho(\alpha) = 0$. We recommend to choose the initial approximation such that the value of the generalized discrepancy $\rho(\alpha)$ at it be positive. If $\|u_\delta\|_{L_2}^2$ exceeds δ^2, this condition will be satisfied for all sufficiently large values of the regularization parameter. If the condition $\rho(\alpha) > 0$ is not satisfied for the initial approximation, we multiply this initial approximation by 2 until the generalized discrepancy becomes positive (but at most IMAX times). Then we construct the Newton sequence, starting with this parameter value. After finishing of PTIKR, the parameter ALFA contains a value of the regularization parameter satisfying the condition $|\rho(\alpha)| \leq \epsilon$.

U is a work array of length at least $6 * N$.

NU is an input parameter; it is the length of the work array U. It serves to control the sufficiency of the array U.

IERR is an output parameter; it is the code with which the program has finished.

IERR $= 0$ if we have found an extremal z^α of $M^\alpha[z]$ for the fixed value

ALFA of the regularization parameter (C1 $\leq$ 1.0), or if we have found by Newton's method a value for the regularization parameter **ALFA** that satisfies the generalized discrepancy principle, if this is what was required (C1 > 1.0).

IERR = 1 if, for C1 > 1.0, during Newton's method we have encountered negative values of the generalized discrepancy $\rho(\alpha)$, i.e. the convexity of the generalized discrepancy is violated. The solution of the discrepancy equation $\rho(\alpha) = 0$ has been found by the chord method.

IERR = 64 if the length of the array **U** does not suffice; the value of **ALFA** has not been changed; the values of all remaining output parameters are not defined.

IERR = 65 if, for C1 > 1.0, we have found a value of the regularization parameter α such that $\rho(\alpha) < 0$. Multiplication of α by 2 has not made it possible to choose a value for which the generalized discrepancy would be positive, i.e. $\rho(\alpha 2^{i_{max}}) < 0$. When PTIKR has finished, **ALFA** contains $2^{i_{max}}$ times the value of the initial regularization parameter, and **Z** contains an extremal corresponding to this parameter.

IERR = 66 if, for C1 > 1.0, we have performed **IMAX** iterations of Newton's method but have not found a root of the discrepancy inequality $|\rho(\alpha)| \leq \epsilon = (c_1 - 1)\delta^2$. When the program has finished, **ALFA** contains the current approximation of the regularization parameter, and **Z** contains an extremal corresponding to it.

IERR = 67 if we have encountered during Newton's method a value of the generalized discrepancy that is less than zero, and have passed to the chord method (which is possible if the generalized discrepancy $\rho(\alpha)$ is not convex ($h \neq 0$)). We have performed in total **IMAX** iterations of Newton's method and the chord method, but have not found a solution of the discrepancy equation. **ALFA** contains a current approximation by the chord method, and **Z** contains an extremal of $M^{\alpha}[z]$ corresponding to it.

When using PTIKR to solve the equation (4.8), it may turn out to be necessary to determine an extremal of $M^{\alpha}[z]$ for various values of α and one and the same righthand side of (4.8). The direct use of PTIKR involves repeatedly taking the Fourier transform of one and the same function. To avoid this repeated transformation, there is the entry PTIKRE in PTIKR. When repeatedly calling for the program PTIKR it suffices to go not to PTIKR itself, but to call the entry PTIKRE with the same actual parameters, having the same meaning. When calling PTIKRE we have to restore before this call the work array U. Note that we can also use PTIKRE when choosing the regularization parameter, e.g. in case PTIKR has finished with code IERR = 65, 66, 67.[2]

The proper solution of the integral equation is done by the program PTIKR1, which implements the search for the extremal of $M^{\alpha}[z]$ and of the solution of the discrepancy

---

[2]We can also use PTIKR to solve equation (4.8) with a periodic kernel. In that case the kernel $K(t)$ of period $T = d - c$ must be given on an interval $[L_1, L_2]$ of length $L_2 - L_1 = 2T$. The boundaries L_1, L_2 must be compatible with the support $[c, d]$ of the righthand side and with the support $[a, b]$ of the solution: $b - a = d - c$. E.g., if $[c, d] = [a, b]$, then we have to choose $L_1 = -T = c - d$ and $L_2 = T$.

equation $\rho(\alpha) = 0$ by Newton's method and the chord method.

In lines 7–20 of the program listing of PTIKR1 we compute that values of the kernel and perform a Fourier transformation on the kernel and the righthand side, by calling the standard program FTF1C written by A.F. Sysoev (see Appendix IV for a program listing).

The program FTF1C forms the real and imaginary parts (called ARE and AIM) of the expression

$$\tilde{z}_m = \sum_{j=1}^{n} \exp\left\{-i\frac{2\pi}{n}(j-1)(m-1)\right\} z_j$$

or

$$\tilde{z}_m = \sum_{j=1}^{n} \exp\left\{i\frac{2\pi}{n}(j-1)(m-1)\right\} z_j$$

for a given vector z, in dependence on its last parameter P. After the program has finished, the real and imaginary parts of the initial vector have been replaced by the real and imaginary parts of its discrete Fourier transform.[3]

When the program PTIKR1 is called for repeatedly, Fourier transformation is not performed.

In lines 21–47 we choose a value of the regularization parameter to which a positive value of the generalized discrepancy corresponds. If the value of the variable EPRO, the accuracy of the solution of the discrepancy equation, is put equal to zero, then the program halts after the first computation of an extremal. In the opposite case, the lines 48–83 implement Newton's method for solving the discrepancy equation. If Newton's method leads to a negative value of the generalized discrepancy, then the program switches to the chord method, which is implemented in lines 84–117.

Finally, the lines 118–148 contain the codes for finishing the program and perform the inverse Fourier transformation, again by calling FTF1C with parameter P < 0.

As test computation we propose to use the solution of the equation

$$\int_0^1 K(x-s)z(s)\,ds = u(x), \qquad x \in (0,2).$$

As kernel we take $K(t) = e^{-80(t-0.5)^2}$, with local support in $(0,1)$. For the exact solution we take

$$\overline{z}(s) = \left(\frac{e^{-(s-0.3)^2/0.03} + e^{-(s-0.7)^2/0.03}}{0.9550408} - 0.052130913\right) * 1.4s.$$

The values of the righthand side $u(x)$ on the uniform grid in the interval $[0,2]$ are determined by the product of the circulant matrix A approximating the integral operator occurring in the equation in accordance with the rectangle formula (Figure 4.5) by the vector of values of the exact solution. We take the errors of specifying the righthand side and the operator equal to $h^2 = 10^{-9}$ and $\delta^2 = 10^{-8}$.

---

[3] We have to note that during the last years programs implementing the fast discrete Fourier transform have been developed with great intensity. If the reader has other standard programs to perform the FFT, which could be more suitable or faster, he/she has to make the corresponding changes in PTIKR1.

Applying the program PTIKR to this problem gives a solution whose numerical values have been listed in Figure 4.5. This solution is an extremal of $M^\alpha[z]$ for regularization parameter $\alpha = 5.7 \cdot 10^{-7}$ and corresponds to the discrepancy value $3.44 \cdot 10^{-8}$.

As initial approximation of the regularization parameter we have used $\alpha = 1$, and the discrepancy equation has been solved with accuracy $0.000001\,\delta^2$, i.e. we have looked for an α such that $|\rho(\alpha)| \leq 0.000001\,\delta^2$.

```
00001          IMPLICIT REAL*8(A-H,O-Z)
00002          IMPLICIT INTEGER*4(I-N)
00003          REAL*8 KERN,L1,L2
00004          DIMENSION U(400),U0(64),Z0(64),Z(64),
               *KERN(64)
00005          EXTERNAL AK
00006          C=0.0
00007          D=2.0
00008          L1=0.0
00009          L2=1.0
00010          ALPHA=1.0
00011          C1=1.000001
00012          N=64
00013          IMAX=500
00014          ST=(D-C)/N
00015          DO 5 I=1,N
00016          S=(I-0.5)*ST-0.5
00017          Z0(I)=0.
00018          IF(S.LT.0.0.OR.S.GT.1.0)GOTO 5
00019          Z0(I)=((DEXP(-(S-0.3)**2/0.03)+
               +DEXP(-(S-0.7)**2/0.03))/0.955040800-
               -0.052130913)*1.4*S
00020     5 CONTINUE
00021       DO 554 K=1,N
00022       S=0.5*(L1+L2)+ST*(K-N/2-1)
00023       KERN(K)=0.
00024       IF(S.LT.L1.OR.S.GT.L2)GOTO 554
00025       KERN(K)=AK(S)
00026   554 CONTINUE
00027       HH=1.E-9
00028       DO 771 I=1,N
00029       S=0.0
00030       DO 770 J=1,N
00031       IND=I-J+N/2+1
00032       IF(IND.LT.1.OR.IND.GT.N)GOTO 770
00033       S=S+KERN(IND)*Z0(J)
00034   770 CONTINUE
00035       U0(I)=S*ST
00036   771 CONTINUE
00037       DL=1.E-8
00038       PRINT 999,(Z0(I),I=1,N)
00039   999 FORMAT(' Exact solution:'/(5F11.8))
00040       PRINT 553,DL,HH
00041   553 FORMAT('  Errors:'/
```

```
               *'            in righthand side:',D16.9/
               *'            in operator        :',D16.9)
00042          CALL PTIKR(AK,U0,A,B,C,D,L1,L2,N,Z,AN,
              *DL,HH,C1,IMAX,ALPHA,U,400,IERR)
00043          PRINT 501,IERR,AN,ALPHA,
              *A,B,(Z(II),II=1,N)
00044     501 FORMAT(' .'/
              *' Return code (IERR)              :',
              *                             I5/
              *' Discrepancy                     :',
              *                             D16.9/
              *' Regularization parameter        :',
              *                             D16.9/
              *' Solution found on interval   : (',
              *             F4.1,',',F3.1,')'/
              *' .'/' Approximate solution:'/
              *                        (5F11.8))
00045          STOP
00046          END
```

```
00001          FUNCTION AK(Y)
00002          IMPLICIT REAL*8(A-H,O-Z)
00003          AK=DEXP(-80.*(Y-.5)**2)
00004          RETURN
00005          END
```

Exact solution:
```
.00000000   .00000000   .00000000   .00000000   .00000000
.00000000   .00000000   .00000000   .00000000   .00000000
.00000000   .00000000   .00000000   .00000000   .00000000
.00000000   .00040570   .00469802   .01649271   .03976953
.07814444   .13325059   .20303119   .28077126   .35567888
.41531472   .44932413   .45318938   .43054079   .39307592
.35804191   .34403504   .36623085   .43211954   .53865959
.67164363   .80785932   .92004465   .98364014   .98334749
.91718612   .79650698   .64202556   .47754936   .32383759
.19461398   .09552648   .02555887   .00000000   .00000000
.00000000   .00000000   .00000000   .00000000   .00000000
.00000000   .00000000   .00000000   .00000000   .00000000
.00000000   .00000000   .00000000   .00000000
```

Errors:
```
        in righthand side:    .999999994D-08
        in operator       :   .999999972D-09
```

```
Return code (IERR)            :   0
Discrepancy                   :   .344029501D-07
Regularization parameter      :   .573252726D-06
Solution found on interval    :   ( -.5,1.5)
```

Approximate solution:
```
-.00017321  -.00009008   .00000358   .00006029   .00008013
 .00007742   .00004968  -.00001771  -.00011536  -.00018162
-.00012615   .00008648   .00034550   .00039636   .00003549
-.00048053   .00007864   .00429205   .01602516   .03971215
 .07899454   .13503747   .20511177   .28210682   .35544513
 .41342703   .44649610   .45048758   .42878790   .39255440
 .35866264   .34570400   .36896886   .43575007   .54236307
 .67392962   .80724469   .91610707   .97765767   .97787558
 .91464292   .79762609   .64528134   .48012463   .32370988
 .19255030   .09573080   .03446567   .00344908  -.00649811
-.00528917  -.00088940   .00212787   .00252328   .00120195
-.00032238  -.00105252  -.00086547  -.00023138   .00029757
 .00044660   .00027443   .00000811  -.00015972
```

FIGURE 4.5

4. Description of the program for solving two-dimensional integral equations of convolution type

In this section we will describe a program for solving two-dimensional first-kind Fredholm integral equations of convolution type

$$Az = \int_{-\infty}^{+\infty} \int_{-\infty}^{+\infty} K(x - s, y - t) z(s, t) \, ds \, dt = u(x, y). \qquad (4.9)$$

Suppose we know that $\operatorname{supp} K(u, w) \subset [l_1, L_1] \times [l_2, L_2]$ while $\operatorname{supp} \overline{z}(s, t) \subset [a, A] \times [b, B]$. Then $\operatorname{supp} \overline{u}(x, y) \subset [c, C] \times [d, D]$, where $c = a + l_1$, $C = A + L_1$, $d = b + l_2$, $D = B + L_2$. We will assume that the approximate righthand side $u_\delta(x, y)$ is known on the rectangle $[c, C] \times [d, D]$. Defining $z(s, t)$ to be equal to zero outside $[a, A] \times [b, B]$, we obtain the equation:

$$Az = \int_{c - \frac{l_1 + L_1}{2}}^{C - \frac{l_1 + L_1}{2}} \int_{d - \frac{l_2 + L_2}{2}}^{D - \frac{l_2 + L_2}{2}} K(x - s, y - t) z(s, t) \, ds \, dt = u(x, y),$$

since

$$[a, A] \times [b, B] \subset \left[c - \frac{l_1 + L_1}{2}, C - \frac{l_1 + L_1}{2} \right] \times \left[d - \frac{l_2 + L_2}{2}, D - \frac{l_2 + L_2}{2} \right].$$

Extending the kernel periodically from its local support $[l_1, L_1] \times [l_2, L_2]$ with period $T_1 = C - c$ in the first and $T_2 = D - d$ in the second argument, we have the possibility of applying the results for solving two-dimensional equations of convolution type using the two-dimensional discrete fast Fourier transform as given in Chapter 1.

The solution algorithm is implemented in the program PTITR. To solve the equation

$$\rho(\alpha) = \| A_h z_\eta^\alpha - u_\delta \|_{L_2}^2 - \left(\delta + h \| z_\eta^\alpha \|_{W_2^1} \right)^2 = 0$$

we have used Newton's method. The derivative $\rho'(\alpha)$ needed in this method can be easily computed in explicit form. All remarks in the previous Section concerning the convergence of this method remain in force.

The program PTITR is the main program, and it is called for as follows:

```
CALL PTITR(AK, UØ, ALIM, N1, N2, Z, DL, H, C1, ALFA, AN2,
* U, NU, IMAX, IERR)
```

Here:

AK(U,W) is the function-subroutine for computing the kernel $K(u, w)$ of (4.9);

UØ is the array of values of the righthand side $u_\delta(x, y)$ on the uniform grids $\{x_k\}_{k=1}^{n_1}$, $\{y_l\}_{l=1}^{n_2}$ in the rectangle $[c, C] \times [d, D]$. The numbers of points in the grids (i.e. n_1 and n_2) must be integral powers of two.

ALIM is a real array of length 12; its last 8 elements contain the values of c, C, d, D, l_1, L_1, l_2, L_2, respectively. The first 4 elements contain the computed values of

$$c - \frac{l_1 + L_1}{2}, \quad C - \frac{l_1 + L_1}{2}, \quad d - \frac{l_2 + L_2}{2}, \quad D - \frac{l_2 + L_2}{2},$$

i.e. the boundaries of the rectangle on which the approximate solution is given.

N1, N2 are the dimensions of the uniform grids in the first and second argument, respectively; these grids, on which the righthand side has been given, have the form

$$x_j = \frac{C-c}{n_1}\left(j - \frac{1}{2}\right) + c, \qquad j = 1, \ldots, n_1,$$

$$y_k = \frac{D-d}{n_2}\left(k - \frac{1}{2}\right) + d, \qquad k = 1, \ldots, n_2.$$

The solution is looked for on grids $\{s_j\}_{j=1}^{n_1}$, $\{t_k\}_{k=1}^{n_2}$ such that

$$s_j = x_j - \frac{l_1 + L_1}{2}, \qquad t_k = y_k - \frac{l_2 + L_2}{2}.$$

The numbers N1 and N2 must be integral powers of two.

Z is an output parameter; it is an array of length N1*N2. When PTITR has finished, Z contains an extremal of $M^\alpha[z]$ on the grids given above for the value of the regularization parameter stored in ALFA.

DL is an input parameter; it is the value of the error δ^2 of specifying the righthand side : $\delta^2 \geq \|u_\delta - u\|_{L_2}^2$.

H is an input parameter; it is the value of the error h^2 of specifying the operator A: $h^2 \geq \|A - A_h\|^2$. If we know instead of the kernel $K(u,w)$ an a function $K_h(u,w)$, given with an error on the rectangle $[l_1, L_1] \times [l_2, L_2]$, then h^2 can be estimated from above:

$$h^2 \leq 4(L_2 - l_2)(L_1 - l_1)\int_{l_1}^{L_1}\int_{l_2}^{L_2}|K(u,w) - K_h(u,w)|^2\,du\,dw.$$

C1 is an input parameter; it determines the mode of functioning of the program:
   C1 $\leq$ 1.0: determine an extremal of $M^\alpha[z]$ for the fixed value ALFA of the regularization parameter;
   C1 $>$ 1.0: choose the regularization parameter in accordance with the generalized discrepancy principle, i.e. from the equation $\rho(\alpha) = 0$; here, the equation has to be solved with accuracy $\epsilon = (c_1 - 1)\delta^2$ (where c_1 is the value of C1), i.e. we look for an α such that $|\rho(\alpha)| \leq \epsilon$.

ALFA is an input parameter in the first mode (C1 $\leq$ 1.0); in the second mode (C1 $>$ 1.0) it is both an input and an output parameter. Its initial value serves as the first approximation for the root of the discrepancy equation.

AN2 is an output parameter. When the program has finished it contains the value of the discrepancy $\|A_h z_\eta^\alpha - u_\delta\|_{L_2}^2$ at the extremal of $M^\alpha[z]$ stored in Z.

IMAX is an input parameter; it is the maximum number of iterations allowed when solving the discrepancy equation $\rho(\alpha) = 0$.

U is a work array of length at least $5 * N1 * N2 + N1 + N2$.

NU is an input parameter; it is the length of the work array U.

IERR is an output parameter; it is the code with which the program has finished.
   IERR = 0 if we have found an extremal z^α of $M^\alpha[z]$ for the fixed value ALFA of the regularization parameter (C1 $\leq$ 1.0), or if we have found by Newton's method a value for the regularization parameter ALFA

that satisfies the generalized discrepancy principle, if this is what was required ($\mathtt{C1} > 1.0$).

$\mathtt{IERR} = 1$ if, for $\mathtt{C1} > 1.0$, during Newton's method we have encountered negative values of the generalized discrepancy $\rho(\alpha)$, i.e. the convexity of the generalized discrepancy is violated. The solution of the discrepancy equation $\rho(\alpha) = 0$ has been found by the chord method.

$\mathtt{IERR} = 64$ if the length of the array $\mathtt{U}$ does not suffice; the value of $\mathtt{ALFA}$ has not been changed; the values of all remaining output parameters are not defined.

$\mathtt{IERR} = 65$ if, for $\mathtt{C1} > 1.0$, we have found a value of the regularization parameter α such that $\rho(\alpha) < 0$. Multiplication of α by 2 has not made it possible to choose a value for which the generalized discrepancy would be positive, i.e. $\rho(\alpha 2^{i_{max}}) < 0$. When $\mathtt{PTITR}$ has finished, $\mathtt{ALFA}$ contains $2^{i_{max}}$ times the value of the initial regularization parameter, and $\mathtt{Z}$ contains an extremal corresponding to this parameter.

$\mathtt{IERR} = 66$ if, for $\mathtt{C1} > 1.0$, we have performed $\mathtt{IMAX}$ iterations of Newton's method but have not found a root of the discrepancy inequality $|\rho(\alpha)| \le \epsilon = (c_1 - 1)\delta^2$. When the program has finished, $\mathtt{ALFA}$ contains the successive approximation of the regularization parameter, and $\mathtt{Z}$ contains an extremal corresponding to it.

$\mathtt{IERR} = 67$ if we have encountered during Newton's method a value of the generalized discrepancy that is less than zero, and have passed to the chord method (which is possible if the generalized discrepancy $\sigma(\mu)$ is not convex ($h \ne 0$)). We have performed in total $\mathtt{IMAX}$ iterations of Newton's method and the chord method, but have not found a solution of the discrepancy equation. $\mathtt{ALFA}$ contains a successive approximation by the chord method, and $\mathtt{Z}$ contains an extremal of $M^{\alpha}[z]$ corresponding to it.

$\mathtt{IERR} = 68$ if, for $\mathtt{C1} > 1.0$, i.e. when choosing the regularization parameter, the initial approximation stored in $\mathtt{ALFA}$ is zero or if the value of $\alpha = 0$ occurred in the process of iteratively solving the discrepancy equation. $\mathtt{Z}$ contains an extremal of $M^0[z]$.

When using $\mathtt{PTITR}$ to solve the equation (4.9), it may turn out to be necessary to determine an extremal of $M^{\alpha}[z]$ for various values of α and one and the same righthand side of (4.9). The direct use of $\mathtt{PTITR}$ involves repeatedly taking the Fourier transform of one and the same function. To avoid this repeated transformation, there is the entry $\mathtt{PTITRE}$ in $\mathtt{PTITR}$. When repeatedly calling for the program $\mathtt{PTITR}$ it suffices to go not to $\mathtt{PTITR}$ itself, but to call the entry $\mathtt{PTITRE}$ with the same actual parameters, having the same meaning. When calling $\mathtt{PTITRE}$ we have to restore before this call the work array $\mathtt{U}$. Note that we can also use $\mathtt{PTITRE}$ when choosing the regularization parameter, e.g. in case $\mathtt{PTITR}$ has finished with code $\mathtt{IERR} = 65, 66, 67$.

The functions and subroutines called for, $\mathtt{PTIKR3}$, $\mathtt{FTFTC}$, $\mathtt{PTICR1}$, $\mathtt{PTICR2}$, are listed in Appendices V and IX. For the program $\mathtt{FTF1C}$ implementing the discrete fast Fourier transform see Appendix IV.

As test example we consider the application of **PTITR** for solving the integral equation

$$\int_0^1 \int_0^1 \exp\left\{-80\left[(x-s-0.5)^2 + (y-t-0.5)^2\right]\right\} z(s,t)\,ds\,dt, \qquad x,y \in (0,2).$$

For the exact solution $\bar{z}(s,t)$ we take

$$\bar{z}(s,t) = \left(\frac{e^{-(s-0.3)^2/0.03} + e^{-(s-0.7)^2/0.03}}{0.955040800} - 0.052130913\right) e^{-(t-0.5)^2/0.03}.$$

We have assumed that the local support of the kernel is concentrated in the rectangle $[0,1] \times [0,1]$.

The call of **PTITR** for solving this equation and the numerical values of the solution for $s = 0.5$, $t = 0.5$ have been given in Figure 4.6.

```
00001          IMPLICIT REAL*8(A-H,O-Z)
00002          IMPLICIT INTEGER*4(I-N)
00003          REAL*8 KERN
00004          DIMENSION U(6500),U0(32,32),Z0(32,32),
               *Z(32,32),ALIM(12),KERN(32,32)
00005          EXTERNAL AK
00006          ALIM(5)=0.
00007          ALIM(6)=2.0
00008          ALIM(7)=0.
00009          ALIM(8)=2.0
00010          ALIM(9)=0.
00011          ALIM(10)=1.0
00012          ALIM(11)=0.0
00013          ALIM(12)=1.0
00014          ALPHA=1.000
00015          C1=1.000001D0
00016          N1=32
00017          N2=32
00018          IMAX=500
00019          R=0.
00020          ST1=(ALIM(6)-ALIM(5))/N1
00021          ST2=(ALIM(8)-ALIM(7))/N2
00022          DO 5 I=1,N1
00023          DO 5 J=1,N2
00024          S1=(I-0.5)*ST1-0.5
00025          S2=(J-0.5)*ST2-0.5
00026          Z0(I,J)=0.
00027          IF(S1.LT.0.0.OR.S1.GT.1.0.OR.S2.LT.0.0.OR.S2.GT.1.0)
               *      GOTO 5
00028          Z0(I,J)=((DEXP(-(S1-0.3)**2/0.03)+
               +DEXP(-(S1-0.7)**2/0.03))/0.9550408-
               -0.052130913)*DEXP(-(S2-0.5)**2/0.03)
00029        5 CONTINUE
00030          DO 554 K=1,N1
00031          DO 554 L=1,N2
00032          S1=(ALIM(9)+ALIM(10))/2.+ST1*(K-N1/2-1)
00033          S2=(ALIM(11)+ALIM(12))/2.+ST2*(L-N2/2-1)
00034          KERN(K,L)=0.
00035          IF(S1.LT.ALIM(9).OR.S1.GT.ALIM(10).
               *   OR.S2.LT.ALIM(11).OR.S2.GT.ALIM(12))
               *                          GOTO 554
00036          KERN(K,L)=AK(S1,S2)
00037      554 CONTINUE
00038          HH=3.68D-11
```

```
00039          DO 771 I1=1,N1
00040          DO 771 I2=1,N2
00041          S=0.
00042          DO 770 J1=1,N1
00043          DO 770 J2=1,N2
00044          IND1=I1-J1+N1/2+1
00045          IND2=I2-J2+N2/2+1
00046          IF(IND1.LT.1.OR.IND1.GT.N1) GOTO 770
00047          IF(IND2.LT.1.OR.IND2.GT.N2) GOTO 770
00048          S=S+KERN(IND1,IND2)*Z0(J1,J2)
00049      770 CONTINUE
00050          U0(I1,I2)=S*ST1*ST2
00051          IF(R.LT.U0(I1,I2)) R=U0(I1,I2)
00052      771 CONTINUE
00053          DL=6.32D-07
00054          PRINT 553,DL,HH
00055      553 FORMAT('   Errors :'/
               *'in righthand side:                 ',D16.9/
               *'        in operator:               ',D16.9)
00056          CALL PTITR(AK,U0,ALIM,N1,N2,Z,DL,HH,
               *C1,ALPHA,AN,U,6500,IMAX,IERR)
00057          PRINT 501,IERR,AN,ALPHA
00058      501 FORMAT(' .',/
               *' Return code               :',I5/
               *' Discrepancy               :',D16.9/
               *' Regularization parameter  :',D16.9)
00059          PRINT 503
00060          PRINT 502,(Z0(13,I),I=1,N2),
               *(Z(13,I),I=1,N2)
00061          PRINT 505
00062          PRINT 502,(Z0(I,16),I=1,N1),
               *(Z(I,16),I=1,N1)
00063      502 FORMAT('      Exact solution:'/8(4F8.4/))
00064      503 FORMAT('.'/'   section X=0.28125   ')
00065      505 FORMAT('.'/'   section Y=0.46875   ')
00066     1000 STOP
00067          END
```

```
00001          FUNCTION AK(X,Y)
00002          IMPLICIT REAL*8(A-H,O-Z)
00003          AK=DEXP(-20.*(X-0.5)**2-20.*(Y-0.5)**2)
00004          RETURN
00005          END
```

Errors :
   in righthand side: .632000000D-06
   in operator: .368000000D-10
.

Return code   :   0
Discrepancy   :   .782989588D-06
Regularization parameter   :   .110838081D-07

section X=0.28125
    Exact solution:
    .0000   .0000   .0000   .0000
    .0000   .0000   .0000   .0000
    .0006   .0040   .0192   .0706
    .2000   .4369   .7354   .9542
    .9542   .7354   .4369   .2000
    .0706   .0192   .0040   .0006
    .0000   .0000   .0000   .0000
    .0000   .0000   .0000   .0000

  -.0019   -.0033   -.0036   -.0004
   .0063   .0120   .0105   -.0009
  -.0155   -.0162   .0194   .1083
   .2493   .4178   .5706   .6616
   .6616   .5706   .4178   .2493
   .1083   .0194   -.0162   -.0155
  -.0009   .0105   .0120   .0063
  -.0004   -.0036   -.0033   -.0019

section Y=0.46875
    Exact solution:
    .0000   .0000   .0000   .0000
    .0000   .0000   .0000   .0000
    .0408   .1950   .4586   .7633
    .9542   .9152   .7023   .5123
    .5123   .7023   .9152   .9542
    .7633   .4586   .1950   .0408
    .0000   .0000   .0000   .0000
    .0000   .0000   .0000   .0000

   .0044   .0074   .0075   -.0018
  -.0200   -.0353   -.0265   .0277
   .1366   .2878   .4498   .5836
   .6616   .6810   .6637   .6431
   .6431   .6637   .6810   .6616
   .5836   .4498   .2878   .1366
   .0277   -.0265   -.0353   -.0200
  -.0018   .0075   .0074   .0044

FIGURE 4.6

5. Description of the program for solving ill-posed problems on special sets. The method of the conditional gradient

The programs considered in this section are intended for the solution of the integral equation (4.1) with exactly specified operator $(h = 0)$. If $h \neq 0$, to solve (4.1) we can use these programs by taking instead of δ, the error of specifying the righthand side, a quantity $\delta + hC_0$, where C_0 is an upper bound for the norm of the exact solution (see Chapter 2, §1). We can readily obtain such an upper bound in each case, basing ourselves on the known constants bounding the exact solution.

The method of the conditional gradient (Chapter 3, §1) for solving (4.1) on the set $Z{\downarrow}_C$ of monotone nonincreasing, the set $\check{Z}{\downarrow}_C$ of monotone nonincreasing upper convex, the set $\check{Z}_C$ of upper convex bounded functions, and also on the set $V_C = \{z: V_a^b(z) \leq C\}$ of functions of bounded variation, has been implemented in the programs PTIGR, PTIGR1, PTIGR2.

Note that problems of solving (4.1) when there is a priori information regarding to inclusion of the exact solution to a set of functions whose direction of monotonicity or convexity does not coincide with those regarded in a program PTIG, can be readily reduced, by a simple change of variables, to a form in which we can immediately use this program.

The algorithm implemented in PTIGR constructs a minimizing sequence for the discrepancy functional

$$\Phi(z) = \int_c^d \left[\int_a^b K(x,s)z(s)\,ds - u(x) \right]^2 dx$$

(of course, using its finite-difference approximation) until the condition $\Phi(z) \leq \delta^2$ is violated, where δ is the error of specifying the righthand side: $\|u_\delta - \overline{u}\|_{L_2} \leq \delta$.

5.1. Description of the program PTIGR.
The program PTIGR is the main program, and it is called for as follows:

```
    CALL PTIGR(AK, U∅, A, B, C, D, N, M, Z, AN2, ITER, DL,
  * IMAX, C1, C2, IC, U, NU, IERR)
```

Here:

AK(X,S) is the function-subroutine for computing the kernel $K(x,s)$ of (4.1).

U∅ is an input parameter; it is the array of values of the righthand side $u_\delta(x)$ on the uniform grid $\{x_i\}_{i=1}^m$ in the interval $[c,d]$, $x_1 = c$, $x_m = d$. The grid consists of M points.

A, B, C, D are input parameters; they are equal to the quantities a, b, c, d, respectively, in (4.1).

N is an input parameter; it is the dimension of the uniform grid $\{s_j\}_{j=1}^n$ in $[a,b]$, $s_1 = a$, $s_n = b$, on which the unknown function $z(s)$ is looked for.

M is an input parameter; it is the dimension of the uniform grid $\{x_i\}_{i=1}^m$ in $[c,d]$ on which the righthand side of (4.1) is given.

Z is an input and an output parameter. When calling PTIGR it contains an initial value of the extremal of the discrepancy functional (it is an array of length N). This initial value must be an admissible point, i.e. must be monotone, convex, or simultaneously monotone and convex, and bounded above by a constant C_2 and below by a constant C_1, or (if the problem is given in V_C) must have bounded generalized variation: $|z_n| + \sum_{i=1}^{n-1} |z_i - z_{i+1}| \leq C_2$. When PTIGR has finished, Z contains the found approximate solution.

AN2 is an output parameter; it contains the value of the discrepancy functional $\Phi(z) = \|Az - u_\delta\|_{L_2}^2$ on the approximate solution found.

ITER is an output parameter; it contains the number of iterations of the method of the conditional gradient performed in the program.

DL is an input parameter; it is the value of the error δ^2 of specifying the righthand side of (4.1): $\delta^2 \geq \|\bar{u}_\delta - u\|_{L_2}^2$. The program PTIGR minimizes $\Phi(z)$ until this value is reached.

IMAX is an input parameter; it is the maximum admissible number of iterations.

C1, C2 are input parameters; they hold the values C_1 and C_2, which bound the solution or its variation.

IC is an input parameter. It determines the correctness set on which PTIGR minimizes $\Phi(z) = \|Az - u_\delta\|_{L_2}^2$.

> IC $= 1$ if this is the set of monotone nonincreasing functions bounded below by C_1 and above by C_2, respectively.
>
> IC $= 2$ if this is the set of monotone nonincreasing upper convex functions bounded below by C_1 and above by C_2, respectively.
>
> IC $= 3$ if this is the set of upper convex functions bounded below by C_1 and above by C_2, respectively. In this case the minimization is actually done on the set $\breve{M}'_{2C} = C\{T^{(j)}\}_{j=0}^n$ which is the convex hull of the vectors

$$T_i^{(0)} = C_1, \qquad T_i^{(j)} = \begin{cases} C_1 + 2(C_2 - C_1)\frac{i-1}{j-1}, & i < j, \\ C_1 + 2(C_2 - C_1), & i = j,\ j = 1,\dots,n, \\ C_1 + 2(C_2 - C_1)\frac{n-i}{n-j}, & i > j. \end{cases} \tag{4.10}$$

> IC $= -1$ if the minimization is done on the set of functions satisfying the condition $V_a^b(z) + z(b) \leq C_2$ (see Chapter 3, §1). The value of C1 must be put equal to zero.

U is a work array of length at least $N * M + 2 * N + 2 * NM$.

NU is an input parameter; it is the length of the work array U.

IERR is an output parameter; it is the code with which the program has finished.

> IERR $= 0$ if we have reached a value of $\Phi(z)$ less than DL or if we have performed IMAX iterations of the method of the conditional gradient.
>
> IERR $= 1$ if at a subsequent step the discrepancy did not decrease; further application of the method of the conditional gradient is not useful. Z will contain the successive approximation and AN2 the discrepancy corresponding to it.
>
> IERR $= 64$ if the length of the array U does not suffice; the values of all output parameters are not defined.

The subroutines called for are: PTIGR1, PTIGR2, PTICR0, PTICRØ, PTICR1, PTICR2, PTICR3, PTICR4, PTICR5, PTICR6.

We will briefly describe the functioning of PTIGR. The lines 7–13 initialize the work arrays and verify the sufficiency of the length of U for the programs rearranging the work arrays. The subsequent call for PTICRØ makes up the matrix A of the operator approximating the integral operator in (4.1). Finally, PTIGR1 actually realizes minimization of the functional $\varphi(z)$ approximating the discrepancy functional $\Phi(z)$ on the sets $M\downarrow_C$, $\check{M}\downarrow_C$, $\check{M}'_{2C}$, or $M_{V_C} = \{z: |z_n| + \sum_{i=1}^{n-1} |z_j - z_{j+1}| \le C\}$.

5.2. Description of the program PTIGR1. Since this program could be of independent interest, we will describe in some detail its functioning. It is intended for minimizing the quadratic function

$$\varphi(z) = \sum_{i=1}^{m} \left(\sum_{j=1}^{n} A_{ij} z_j - u_i \right)^2 \tag{4.11}$$

by the method of the conditional gradient on certain special sets.

The program is called for as follows:

    CALL PTIGR1(A, ZØ, UØ, C1, C2, IC, N, M, ITER, DL,
    * ANGRD, IMAX, AN2, Z, U, U1, H, G, IERR)

Here:

A is the matrix of the operator $A:\ : \mathbf{R}^n \to \mathbf{R}^m$ occurring in the expression for minimizing $\varphi(z)$ in (4.11).

ZØ is an initial approximation (an array of length N); the initial approximation must be an admissible point.

UØ is the array of length M containing the coordinates of the vector u in (4.11).

C1, C2, IC are parameters determining the set on which $\varphi(z)$ is minimized (C_1, C_2 are the values of C1, C2).

    IC = 1: $\{z: z_i \ge z_{i+1}, C_2 \ge z_i \ge C_1, i = 1,\dots,n\}$.

    IC = 2: $\{z: z_{i-1} - 2z_i + z_{i+1} \le 0, C_2 \ge z_i \ge C_1\}$.

    IC = 3: $C\{T^{(j)}\}_{j=1}^n$, where $T^{(j)}$ are defined by (4.10).

    IC = −1: C1 = 0, $\{z: |z_n| + \sum_{i=1}^{n-1} |z_i - z_{i+1}| \le C_2$.

N, M are the parameters n, m in (4.11).

ITER is the number of iterations performed in the program PTIGR1.

DL is the level with which PTIGR1 minimizes $\varphi(z)$.

ANGRD is the exit level for the square of the norm of the gradient. If the square of the norm of the gradient, $\|\operatorname{grad}\varphi(z)\|^2 = 4\|A^*Az - A^*u\|^2_{\mathbf{R}^m}$, becomes less than this value, the program finishes;

IMAX is a restriction on the number of iterations; PTIGR1 cannot perform more than IMAX iterations.

AN2 is the value of the functional $\varphi(z) = \|Az - u\|^2_{\mathbf{R}^m}$ on the approximate solution found.

Z contains the found approximate solution. When calling PTIGR1 the actual parameters corresponding to Z and ZØ may coincide.

U is the value of Az on the approximate solution found.

G, H are work arrays of length N.

U1 is a work array of length M.

IERR is the code with which the program has finished.

IERR $= 0$ if the program has finished regularly; the level of the discrepancy is reached or IMAX iterations have been performed.

IERR $= 1$ if at a subsequent step the discrepancy did not decrease.

The lines 10–15 successively perform the following actions, calling the appropriate subroutines:

1. The initial approximation Z0 is sent to Z.
2. The product of A and Z is computed and stored in U.
3. The discrepancy $\varphi(z) = \|Az - u\|^2_{\mathbf{R}^m}$ is computed at the initial approximation.
4. The gradient vector $2(A^*Az - A^*u) = g$ of the discrepancy functional at the initial approximation is computed.
5. The square of the norm of the gradient, $\|g\|^2$, is computed, and then the condition for ending the program is verified. If it is not fulfilled, the following iteration process is started:
   a) a call to PTIGR2 computes a vertex H of the polyhedron determined by the parameter IC at which the linear functional $(g, t) = \psi(t)$ attains its minimum;
   b) we compute the vector pointing from the vertex H in the direction of the current solution, i.e. the direction opposite the direction of descent;
   c) PTICR0 minimizes $\varphi(z)$ on the line segment $z - \lambda h$, with $\lambda \in [0, 1]$. The result of this program call is the stepsize λ;
   d) we change the approximate solution in accordance with the stepsize λ found;
   e) we again compute the value of the operator on the new approximation, the gradient at the new point, the value of the discrepancy, and the square of the norm of the gradient;
   f) we verify the condition for ending the program. If it is not satisfied, we verify whether the value of $\varphi(z)$ has decreased; if not then the program ends with code 1; in the opposite case we store the current value of the discrepancy and continue with the iteration process.

5.3. Description of the program PTIGR2. This program is intended for choosing an optimal vertex t_0 of the corresponding polyhedron M_C, i.e. a vertex at which the functional (g, t) is minimal:

$$\psi(t_0) = (g, t_0) = \min_{t \in M_C} \psi(t) = \min_{t \in M_C} (g, t),$$

where g is a given fixed vector.

The program is called for as follows:

    CALL PTIGR2(TOP, G, N, C1, C2, IC)

Here:

TOP is result of the program: the optimal vertex found (it is an array of length N).

G is the gradient g of the discrepancy functional: a fixed given vector of length N.

C1, C2 are the constants C_1, C_2 determining M_C.

IC is a parameter defining M_C (see PTIGR1).

We have implemented the search for the optimal vertex t_0 among the vertices $T^{(j)}$ ($j = 0, \ldots, n$) by simply checking all of them. The coordinates of $T^{(j)}$ can be readily found in explicit form, as done in Chapter 3, §3. When checking optimality we do not have to compute explicitly each time the coordinates of the $T^{(j)}$ ($j = 0, \ldots, n$) and the scalar product $(g, T^{(j)})$. Note that it suffices to compare the scalar products $(g, T^{(j)} - T^{(0)})$. The coordinates of $T^{(j)} - T^{(0)}$ are described in (2.5), (2.8), and (2.11), and (3.7).

As test computation we propose to solve equation (4.1) with kernel $K(x, s)$ defined by (4.2) and intervals $[a, b] = [0, 1]$, $[c, d] = [-2, 2]$. We will use the exact solution $\bar{z}(s) = 1 - s^2$, and when solving the ill-posed problem we will use the a priori information that this solution is monotone and convex. We take the dimensions of the grids in both variables equal to 41 ($n = m = 41$). As righthand side of (4.1) we take the values computed in accordance with (4.3); thus, we have an exactly specified righthand side, and error value equal to zero.

Applying PTIGR to solve this equation, we find an approximate solution $z(s)$ whose numerical values are given in Figure 4.7, which also contains the call of PTIGR. The value of the discrepancy $\Phi(z) = \|Az - u_\delta\|^2$ on the approximate solution found is $2.08 \cdot 10^{-7}$, which corresponds to a relative error of approximately 0.15% (in relation to the maximum of the righthand side). To find the solution given we needed 400 iterations in total.

```
00001          IMPLICIT REAL*8(A-H,O-Z)
00002          IMPLICIT INTEGER*4(I-N)
00003          DIMENSION U0(41),Z(41),U(41,260),Z0(41)
00004          EXTERNAL AK
00005          X1=0.
00006          X2=1.
00007          Y1=-2.
00008          Y2=2.
00009          N=41
00010          M=41
00011          IMAX=400
00012          IC=2
00013          DL=0.
00014          C1=0.
00015          C2=1.
00016          DO 1 J=1,N
00017          X=X1+(X2-X1)/(N-1.)*(J-1.)
00018        1 Z0(J)=1.-X*X
00019          CALL PTICR0(AK,U,X1,X2,Y1,Y2,N,M)
00020          CALL PTICR3(U,Z0,U0,N,M)
00021          DO 4 J=1,N
00022          Z(J)=0.
00023        4 CONTINUE
00024          CALL PTIGR(AK,U0,X1,X2,Y1,Y2,N,M,Z,AN,
               *ITER,DL,IMAX,C1,C2,IC,U,41*260,IERR)
00025          PRINT 5,Z0,Z,DL,AN,ITER,IERR
00026          STOP
00027        5 FORMAT(15X,'Exact solution:'/'.'/
               *                8(5F11.6/),F11.6/'.'/
               *15X,'Approximate solution:'/'.'/
               *                8(5F11.6/),F11.6/'.'/
               *10X,'Square of the error of the righthand side
               *                          D14.6/
               *10X,'Discrepancy                    :',
               *                          D14.6/
               *10X,'Number of iterations           :',
               *                          I14/
               *10X,'Return code (IERR)             :',
               *                          I14)
00028          END
```

```
00001          FUNCTION AK(X,Y)
00002          IMPLICIT REAL*8(A-H,O-Z)
00003          AK=1./(1.+100.*(X-Y)**2)
00004          RETURN
00005          END
```

Exact solution:
```
 1.000000    .999375    .997500    .994375    .990000
  .984375    .977500    .969375    .960000    .949375
  .937500    .924375    .910000    .894375    .877500
  .859375    .840000    .819375    .797500    .774375
  .750000    .724375    .697500    .669375    .640000
  .609375    .577500    .544375    .510000    .474375
  .437500    .399375    .360000    .319375    .277500
  .234375    .190000    .144375    .097500    .049375
  .000000
```

Approximate solution:
```
  .988767    .988767    .988576    .987790    .986477
  .984712    .981902    .974724    .963316    .951637
  .939958    .928279    .916600    .896694    .876788
  .856882    .836975    .814705    .791996    .769288
  .746579    .723870    .701114    .676169    .647221
  .616290    .581252    .545292    .508441    .469801
  .430246    .390691    .351135    .311580    .272024
  .232469    .192914    .150570    .108227    .065883
  .023540
```

Square of the error of the righthand side: .000000D+00
Discrepancy   :   .207510D-06
Number of iterations   :   400
Return code (IERR)   :   0

FIGURE 4.7

6. Description of the program for solving ill-posed problems on special sets. The method of projection of conjugate gradients

The method described in Chapter 3, §2 for solving an integral equation (4.1) on the set $Z\downarrow_C$ of monotone bounded, the set $\check{Z}\downarrow_C$ of monotone convex, and the set of convex nonnegative functions, and based on using the method of projection of conjugate gradients for minimizing the discrepancy functional $\Phi(z) = \|Az - u_\delta\|^2_{L_2}$ on these sets, has been implemented in the programs PTILR, PTILR1, PILR∅, PTILR3, PTILR5, PTILR6, PTILR7, PTILRA, PTILRB.

6.1. Description of the program PTILR. The program PTILR is the main program, and it is called for as follows:

```
CALL PTILR(AK, U∅, A, B, C, D, N, M, Z, AN2, DL, ITER,
* IMAX, C2, IC, U, NU, IERR)
```

Here:

AK(X,S) is the function-subroutine for computing the kernel $K(x,s)$ of (4.1).

U∅ is an input parameter; it is the array of values of the righthand side $u_\delta(x)$ of (4.1) on the uniform grid $\{x_i\}_{i=1}^m$ in the interval $[c,d]$, $x_1 = c$, $x_m = d$. The grid consists of M points.

A, B, C, D are input parameters; they are equal to the quantities a, b, c, d, respectively, in (4.1).

N is an input parameter; it is the dimension of the uniform grid $\{s_j\}_{j=1}^n$ in $[a,b]$, $s_1 = a$, $s_n = b$, on which the unknown function $z(s)$ is looked for.

M is an input parameter; it is the dimension of the uniform grid $\{x_i\}_{i=1}^m$ in $[c,d]$ on which the righthand side of (4.1) is given.

Z is an input and an output parameter. When calling PTILR it contains an initial approximation (it is an array of length N). This initial value must be an admissible point, i.e. must be nonnegative and monotone, convex, or simultaneously monotone and convex, and, if we are dealing with $Z\downarrow_C$ (IC = 1), it must be bounded above by a constant C_2. When PTILR has finished, Z contains the found approximate solution of (4.1).

AN2 is an output parameter; it contains the value of the discrepancy functional $\Phi(z) = \|Az - u_\delta\|^2$ on the approximate solution found.

DL is an input parameter; it is the value of the error δ^2 of specifying the righthand side of (4.1): $\delta^2 \geq \|u_\delta - \bar{u}\|^2$. The program PTILR minimizes $\Phi(z)$ until this value is reached.

ITER is an output parameter; it contains the number of loops of the program (the number of changes in the active constraints).

IMAX is an input parameter; it is the maximum admissible number of loops.

C2 is an input parameter; it is a constant holding the value C_2, which bounds the solution from above, in case we are dealing with $Z\downarrow_C$ (IC = 1).

IC is an input parameter. It determines the set on which PTILR minimizes the discrepancy functional.

> IC = 1 if this is the set of nonnegative, monotonically decreasing functions bounded above by C_2 ($Z\downarrow_C$).

IC = 2 if this is the set of nonnegative upper convex functions ($\check{Z}$).

IC = 3 if this is the set of nonnegative, monotonically decreasing upper convex functions ($\check{Z}\downarrow$).

U is a work array of length at least $nm + 5n^2 + 5n + \max\{n, m\} + 1$. If IC = 1, then its length should be at least $nm + 5n^2 + 8n + \max\{n, m\} + 4$.

NU is an input parameter; it is the length of the work array U.

IERR is an output parameter; it is the code with which the program has finished.

IERR = 0 if we have a minimum of the discrepancy functional $\Phi(z)$ on the corresponding set (we pass to step 6 in the algorithm described in Chapter 3, §2).

IERR = 1 if the attained value of the discrepancy functional is less than the given error value DL.

IERR = 2 if we have found a minimum of the discrepancy functional $\Phi(z)$ (the norm of the gradient of the functional equals the machine zero).

IERR = 3 if we have performed IMAX loops (changes of the set of active constraints).

IERR = 64 if the length of the array U does not suffice; the values of all output parameters are not defined.

IERR = 65 if there is an error in specifying the correctness set; IC is either less than one or larger than three, or we have set C2 = 0.0 while IC = 1; the values of all output parameters are not defined.

The subroutines called for are: PTILR1, PTILR2, PTILR3, PTILR4, PTILR5, PTILR6, PTILR7, PTILRA, PTILRB, PTICRØ, PTICR1, PTICR2, PTICI2, PTICR3, PTICR5, PTICR6, PTILRØ.

We will briefly describe the functioning of PTILR. The lines 9–24 initialize the work arrays and verify the sufficiency of the length of U for the programs rearranging the work arrays. The subsequent call for PTICRØ makes up the matrix of the operator $\mathbf{R}^n \to \mathbf{R}^m$ approximating the integral operator in (4.1). Then, PTILRA rewrites the discrepancy functional $\varphi(z) = \|Az - u_\delta\|^2$ as $\varphi(z) = (z, Qz) + (d, z) + e$, i.e. makes up the matrix $Q = A^*A$ and the vector $d = 2A^*u_\delta$. The program PTILRB makes up the matrix F and vector g of the constraint $Fz \leq g$, in accordance with the value of IC, which determines the correctness set.

The program PTILR1 realizes the main method, attempting to minimize the discrepancy functional $(z, Qz) + (d, z) + e$ in the presence of constraints of the form $Fz \leq g$. The discrepancy AN2, provided by PTILR1, is computed in $\mathbf{R}^m$, and hence after calling PTILR1 it is normalized using the metric of $L_2[c, d]$.

The entry PTILRE, allowing continuation of the computations in the program, makes it possible to avoid repeated calls to the algorithms for constructing the matrix of the operator, the constraints, and the transformation of the discrepancy functional to the form $(z, Qz) + (d, z) + e$. In this case, instead of calling PTILR1 we call its entry PTILR2.

6.2. Description of the program PTILR1.

Since, as already noted, this program implements the main algorithm, it may be of independent interest and we will briefly describe it. It serves to minimize the quadratic function $\|Az - u\|^2_{\mathbf{R}^m} = (z, Qz) +$

$(d, z) + e$ in the presence of constraints of the form $Fz \le g$. Minimization is done using the method of projection of conjugate gradients, as described in Chapter 3, §2.

The program is called for as follows:

```
CALL PTILR1(M, MF, MASK, A, AJ, F, P, UØ, PI, G, GR, W, P1,
* WORK, IMAX, ITER, Q, X, D, N, NN, AN2, DL, ANGRD, IEND)
```

Here:

M is the number of points in the grid on which the righthand side u is given (if WORK $=$.FALSE. it need not be specified).

MF is an integral work variable; it is the number of active constraints.

MASK is a work array of length NN (the number of constraints); it contains indicators for activity of the constraints.

A is the (M $*$ N)-matrix of the operator A (if WORK $=$.FALSE. it need not be specified).

AJ is a work array of length at least NN $*$ N.

F is the (NN $*$ N)-matrix of constraints.

P is a work array of length at least NN $*$ N.

UØ is the righthand side of (4.1) (if WORK $=$.FALSE. it need not be specified).

PI is a work array of length at least N $*$ N.

G is the vector g of constraints; its length is NN.

GR is a work array of length at least N.

W is a work array of length at least max$\{$NN,M,N$\}$.

P1 is a work array of length at least N.

WORK is a local variable, determining the mode of computation of the discrepancy

WORK $=$ .TRUE. the input parameters M, A, UØ and DL have been given. In each loop we compute the discrepancy $\varphi(z) = \|Az - u\|^2_{\mathbf{R}^m}$ and compare it with DL.

WORK $=$ .FALSE. the discrepancy is not computed (the quantities M, A, UØ and DL need not be given when calling the program).

IMAX is the maximum admissible number of loops;

ITER is the number of loops performed.

Q is the matrix Q in the discrepancy functional $\varphi(z) = (z, Qz) + (d, z) + e$.

X is an initial approximation. When the program has finished, the array X o length N contains the approximate solution found.

D is the vector d in the discrepancy functional $\varphi(z) = (z, Qz) + (d, z) + e$. Its length is N.

N is the dimension of the solution space.

NN is the number of constraints.

AN2 contains, when the program has finished, the value of the discrepancy functional $\varphi(z) = \|Az - u\|^2_{\mathbf{R}^m}$ on the approximate solution found, provided that WORK $=$ .TRUE..

DL is the level with which the discrepancy is minimized. If $\varphi(z) = \|Az - u\|^2$ becomes less than DL, the program finishes.

ANGRD is the exit level with respect to the gradient $\| \operatorname{grad} \varphi(z)\|^2 = \|2Qz + d\|^2$ If the square of the norm of the gradient becomes less than this value, the program finishes.

IEND is the code with which the program has finished.

IEND = 0 if we have found a minimum point of $\varphi(z)$ on the set under consideration (we pass to step 6 in the algorithm of Chapter 3, §2).

IEND = 1 if the value of $\varphi(z)$ reached is less than DL.

IEND = 2 if the value of the norm of the gradient reached is less than ANGRD.

IEND = 3 if we have performed IMAX iterations.

The program has an entry PTILR2, intended for the repeated call to continue minimization. In such a case it is necessary to call PTILR2, after having defined all required input parameters and also having restored the values of MF and MASK. We then succeed in avoiding the procedure of accumulating active constraints using fictitious steps requiring the inversion of the matrix.

We draw attention to the fact that the parameters DL and AN2, which characterize the discrepancy, have a different meaning in PTILR and PTILR1.

We can also use PTILR1 when solving (4.1) in the presence of information on the location of, e.g., the extrema of the unknown function. For this it suffices to use in PTILR another program, PTILRB, for making up the matrix F and the constraint vector g. Note that we always have to require that at all admissible points the columns of the matrix of active constraints are independent. In the simplest case it suffices to change only the signs of the corresponding columns of the matrix of constraints. We give as example the program PTILRB for the case when the problem involves the set of functions that are nondecreasing for $a \leq s \leq x_0$ and nonincreasing for $x_0 \leq s \leq b$. We assume that x_0 belongs to the grid on which the solution is looked for, and has been given in a COMMON-block with reference number J∅.

```
      SUBROUTINE PTILRB(F, G, N, NN, ITASK, C, IERR)
      COMMON JO
C * ITASK, C - fictitious parameters
      DIMENSION F(NN, N), G(NN)
      REAL *8 F, G, C
      INTEGER N, NN, ITASK, IERR, L, I, JO
C * make up the number of constraints - NN
      NN = N - 1
C * fill matrix with zeros
      CALL PTICR2(F, 0.0, N * NN)
C * make up the constraint vector
      CALL PTICR2(G, 0.0, NN)
C * make up the constraint matrix
      DO 11 I=2,JO
      F(I-1,I-1)=1.0
      F(I-1,I)=-1.0
11    CONTINUE
      L=N-1
```

```
      DO 12 I=J0,L
      F(I,I)=-1.0
      F(I,I+1)=1.0
12    CONTINUE
      IERR=0
      RETURN
      END.
```

As test computation we propose to solve equation (4.1) with kernel defined by (4.2) ($[a, b] = [0, 1]$, $[c, d] = [-2, 2]$) and exact solution $\overline{z}(s) = 1 - s^2$ by using the program PTILR. We take n, m equal to 41. As righthand side of (4.1) we take the values computed in accordance with (4.3). In Figure 4.8 we have given the program call implementing the solution of this problem, with the use of a priori information regarding the monotone decrease of the unknown function as well as its boundedness above by the constant $C_2 = 1$. We have also listed the results there, i.e. the approximate solution $z(s)$.

Note that when solving (4.1) by using PTILR we have to invert the matrices of active constraints more than once, which leads to a considerable expenditure of computer time. Moreover, specifying an 'unlucky' initial approximation (on an edge of low dimension) may necessarily lead to an accumulation of active constraints, requiring fictitious steps at each of which the matrix of active constraints has to be inverted. Thus, for problems having large dimension the use of PTILR may turn out to be inefficient.

```
00001            IMPLICIT REAL*8(A-H,O-Z)
00002            IMPLICIT INTEGER*4(I-N)
00003            DIMENSION U0(41),Z(41),U(41,260),Z0(41)
00004            EXTERNAL AK
00005            X1=0.
00006            X2=1.
00007            Y1=-2.
00008            Y2=2.
00009            N=41
00010            M=41
00011            IMAX=100
00012            IC=1
00013            DL=0.
00014            C2=1.
00015            DO 1 J=1,N
00016            X=X1+(X2-X1)/(N-1.)*(J-1.)
00017          1 Z0(J)=1.-X*X
00018            CALL PTICR0(AK,U,X1,X2,Y1,Y2,N,M)
00019            CALL PTICR3(U,Z0,U0,N,M)
00020            DO 4 J=1,N
00021            Z(J)=0.5
00022          4 CONTINUE
00023            CALL PTILR(AK,U0,X1,X2,Y1,Y2,N,M,Z,AN,
                *DL,ITER,IMAX,C2,IC,U,41*260,IERR)
00024            PRINT 5,Z0,Z,DL,AN,ITER,IERR
00025            STOP
00026          5 FORMAT(15X,'Exact solution:'/'.'/
                *                8(5F11.6/),F11.6/'.'/
                *15X,'Approximate solution:'/'.'/
                *                8(5F11.6/),F11.6/'.'/
                *10X,'Square of the error of the righthand side
                *                                D14.6/
                *10X,'Discrepancy                      :',
                *                                D14.6/
                *10X,'Number of iterations             :',
                *                                I14/
                *10X,'Return code (IERR)               :',
                *                                I14)
00027            END
```

```
00001          FUNCTION AK(X,Y)
00002          IMPLICIT REAL*8(A-H,O-Z)
00003          IMPLICIT INTEGER*4(I-N)
00004          AK=1./(1.+100.*(X-Y)**2)
00005          RETURN
00006          END
```

Exact solution:
```
1.000000    .999375    .997500    .994375    .990000
 .984375    .977500    .969375    .960000    .949375
 .937500    .924375    .910000    .894375    .877500
 .859375    .840000    .819375    .797500    .774375
 .750000    .724375    .697500    .669375    .640000
 .609375    .577500    .544375    .510000    .474375
 .437500    .399375    .360000    .319375    .277500
 .234375    .190000    .144375    .097500    .049375
 .000000
```

Approximate solution:
```
1.000000    .999382    .997406    .994848    .989138
 .983531    .983531    .961528    .961528    .948344
 .948344    .912060    .912060    .894492    .894492
 .835698    .835698    .835698    .808443    .752639
 .752639    .728142    .728142    .628439    .628439
 .609002    .584398    .584398    .584398    .386254
 .386254    .386254    .386254    .361675    .307676
 .183705    .183705    .159545    .095421    .048805
 .000000
```

```
Square of the error of the righthand side:      .000000D+00
Discrepancy                              :      .442992D-15
Number of iterations                     :          100
Return code (IERR)                       :            3
```

FIGURE 4.8

7. Description of the program for solving ill-posed problems on special sets. The method of conjugate gradients with projection into the set of vectors with nonnegative components

The method described in Chapter 3, §3 for solving an integral equation (4.1) on the set $Z\downarrow_C$ of monotone or the set $\breve{Z}$ of nonnegative functions is based on transition to new variables. It has been implemented in the programs PTISR, PTISR1, PTISR2, PTISR3, PTISR4.

7.1. Description of the program PTISR. The program PTISR is the main program, and it is called for as follows:

        CALL PTISR(AK, U∅, A, B, C, D, N, M, Z, AN2, ITER,
    * DL, IMAX, IC, U, NU, IERR)

Here:

AK(X,S) is the function-subroutine for computing the kernel $K(x,s)$ of (4.1).

U∅ is an input parameter; it is the array of values of the righthand side of (4.1) on the uniform grid $\{x_i\}_{i=1}^m$ in the interval $[c,d]$, $x_1 = c$, $x_m = d$. The grid consists of M points.

A, B, C, D are input parameters; they are equal to the quantities a, b, c, d, respectively, in (4.1).

N is an input parameter; it is the dimension n of the uniform grid $\{s_j\}_{j=1}^n$ in $[a,b]$, $s_1 = a$, $s_n = b$, on which the unknown function $z(s)$ is looked for.

M is an input parameter; it is the dimension of the uniform grid $\{x_i\}_{i=1}^m$ on which the righthand side of (4.1) is given.

Z is an input and an output parameter. When calling PTISR it must contain an initial approximation of the extremal of the discrepancy functional (it is an array of length N). This initial value must be an admissible point, i.e. must be monotone, convex, or simultaneously monotone and convex, and also nonnegative. When PTISR has finished, Z contains the found approximate solution of (4.1).

AN2 is an output parameter; it contains the value of the discrepancy functional $\Phi(z) = \|Az - u_\delta\|_{L_2}^2$ on the approximate solution found.

ITER is an output parameter; it contains the number of iterations performed in PTISR by the method of projection of conjugate gradients.

DL is an input parameter; it is the value of the error δ^2 of specifying the righthand side of (4.1). The program PTISR minimizes $\Phi(z)$ until this value is reached.

IMAX is an input parameter; it is the maximum admissible number of iterations.

IC is an input parameter. It determines the set on which the discrepancy functional is minimized.

> IC = 1 if this is the set of nonnegative, monotonically nonincreasing functions.
>
> IC = 2 if this is the set of nonnegative, monotonically nonincreasing upper convex functions.
>
> IC = 3 if this is the set of nonnegative upper convex functions.

IC = 4 if this is the set of nonnegative, monotonically nonincreasing lower convex functions.

IC = 5 if this is the set of lower convex functions that are nonnegative at the boundaries of the interval under consideration (at a and b).

U is a work array of length at least $N * M + 3 * N + 2 * M$.

NU is an input parameter; it is the length of the work array U.

IERR is an output parameter; it is the code with which the program has finished.

IERR = 0 if we have a minimum of the discrepancy functional $\Phi(z) = \|Az - u_\delta\|^2$ on the corresponding set (the norm of the gradient of the functional equals the machine zero, or we have found a minimum in the current subspace and no constraint can be excluded from the list of active constraints).

IERR = 1 if the attained value of the discrepancy functional is less than the given error value DL, or if we have performed IMAX iterations by the method of projection of conjugate gradients.

IERR = 2 if the discrepancy functional $\Phi(z)$ has not decreased at the subsequent step, and further application of the method of projection of conjugate gradients is not successful. When PTISR has finished, Z contains the successive approximation to the minimum point.

IERR = 64 if the length of the work array does not suffice; the values of all output parameters are not defined.

IERR = 65 if accumulation of rounding errors occurs; the values of all output parameters are not defined.

The subroutines called for are: PTISR1, PTISR2, PTISR3, PTISR4, PTICR0, PTICRØ, PTICR1, PTICR2, PTICI2, PTICR3, PTICR4, PTICR5, PTICR6, PTICR7, PTICR8, PTICR9.

We will briefly describe the functioning of PTISR. The lines 9–16 initialize the work arrays and verify the sufficiency of the length of U for the program PTISR. The subsequent call for PTISRØ makes up the matrix A of the linear operator approximating the integral operator in (4.1). Then, PTISR2 transforms A and the initial approximation Z in accordance with the change of variables mapping the minimization problem into the set Π^+ (see Chapter 3, §3). The program PTISR1 implements the minimization of the discrepancy functional $\varphi(z) = \|Az - u_\delta\|^2_{\mathbf{R}^n}$ on the set Π^+ of vectors with nonnegative components by the method of projection of conjugate gradients. Note that in PTISR the discrepancy is computed in an analog of the norm of $L_2[c, d]$, while in PTISR1 this quantity is computed in the usual Euclidean metric of $\mathbf{R}^n$. When the minimization has finished, the extremal point x is subjected to the inverse transformation $z = Tx$ from Π^+ into the set determined by the value of IC. The program PTISR3 implements this transformation.

The entry PTISRE serves for repeated calls of the program and for continuation of the computations. In this case we do not have to make up A anew and realize its transformation and the transformation of the initial approximation. The matrix A in transformed form is stored in the first $N * M$ elements of the work array U. To store the current approximation $x \in \Pi^+$ we use line 24, which sends this element to the following N elements of U. Thus, to continue the computation it suffices to store and restore the first $N * (M + 1)$ elements of U.

7.2. Description of the program PTISR1. This program is the main implementation of the algorithm. In the general case it minimizes the smoothing functional

$$M^\alpha[z] = \|Az - u\|^2_{\mathbf{R}^m} + \alpha \left(\|z\|^2_{\mathbf{R}^n} + \rho\|z'\|^2_{\mathbf{R}^{n-1}} \right)$$

on the set Π^+ of vectors with nonnegative components $z_i \geq 0$ (if imbedded in PTISR, PTISR1 is used for minimizing the discrepancy, i.e. $\alpha = 0$) by the method of projection of conjugate gradients.

The program is called for as follows:

CALL PTISR1(A, Z∅, U∅, N, M, ITER, DL, ANGRD, IMAX, AN2,
* ALF, RO, Z, U, U1, H, G, IPLUS, IERR)

Here:

A is the matrix of the operator $A: : \mathbf{R}^n \to \mathbf{R}^m$; it is an array of length $M * N$.

Z∅ is an initial approximation (an array of length N). All components of the initial approximation must be nonnegative.

U∅ is the array of length M containing the righthand side of the equation.

N, M are the dimensions of the spaces.

ITER is the number of iterations performed in the program.

DL is the exit level with respect to the discrepancy $\varphi(z) = \|Az - u\|^2_{\mathbf{R}^m}$. If $\varphi(z)$ becomes less than this value, the program finishes with code IERR = 1.

ANGRD is the exit level for the gradient. If $\| \operatorname{grad} M^\alpha[z]\|^2_{\mathbf{R}^n}$ becomes less than this value, the program finishes with IERR = 1.

IMAX is the maximum admissible number of iterations.

AN2 is the value of $\varphi(z)$ on the approximate solution found.

ALF is the value of the regularization parameter in $M^\alpha[z]$.

RO is the value of weight ρ in the expression for $M^\alpha[z]$.

Z contains the found approximate solution. It is an array of length N.

U is the value of $Az \in \mathbf{R}^m$ on the approximate solution found. It is an array of length M.

U1 is a work array of length M.

G, H, IPLUS are work arrays of length N.

IERR is the code with which the program has finished.

> IERR = 0 if we have found an 'exact' minimum of $M^\alpha[z]$ on Π^+.
>
> IERR = 1 if we have performed IMAX iterations, of if we have reached the discrepancy value DL, or if we have attained a value of the square of the norm of the gradient which is less than ANGRD.
>
> IERR = 65 if the computed step along the direction of descent is negative (this is possible only if there is an accumulation of rounding errors).

We will briefly describe the functioning of the program PTISR1. The lines 12–18 successively perform the following actions, calling the appropriate subroutines:

1. The initial approximation is sent to Z.
2. The product of A and Z is computed and stored in U.
3. The discrepancy is computed and the value of the stabilizer is added to it.

4. The gradient of the discrepancy is computed and the gradient of the stabilizer is added to it.

In lines 22–40 we form (or change, in dependence on the value of the flag JHC) the set of active constraints, and compute the dimension of the current boundary.

We then call PTICR7 to compute the square of the norm of the projection of the gradient onto the set of active constraints (on the current face). In lines 42–51 we change and make up the iteration count on the current face, and determine whether the exact minimum is attained on this face. If it is, the set of active constraints is made up anew, and if its dimension does not change by this, we compute whether an exact minimum is attained; in the opposite situation we continue with line 51. The lines 52–56 make up the direction of descent in accordance with the value of the flag ICH for transition to a new face. Then the lines 57–67 get the largest step not leading outside Π^+, and the number of the new possible constraint. In succession the following is done: one-dimensional minimization using PTICR0, transition to a new point, computation of the discrepancy, smoothing functional and their gradients, establishing the necessary flags and verifying the level for finishing the program. After storing the equation for the discrepancy the program returns to line 19, the start of the iteration.

The program PTISR2 serves to transform the matrix of the operator $A \colon \mathbf{R}^n \to \mathbf{R}^m$ and the initial approximation under transition to Π^+.

The program is called for as follows:

CALL PTISR2(A, Z, N, M, IC, S)

Here:

A is the matrix of the operator.

Z is an initial approximation.

N, M are the dimensions of the spaces.

IC is a parameter determining the correctness set M, and has the same meaning as when calling PTISR.

S is a work array of length N.

The program PTISR3 serves to return from Π^+ to the correctness set determined by the parameter IC.

The program is called for as follows:

CALL PTISR3(Z, N, IC, Y)

Here:

Z is the transformed vector of length N.

Y is a work array of length N.

The programs PTISR2 and PTISR3 use the function-subroutine PTISR4, which computes the entries $T_i^{(j)}$ of the transition matrix from Π^+ into the set determined by IC (see Chapter 3, §3). It is called as follows:

S=PTISR4(I, J, IC, N).

The programs described in this section can be modified to solve equation (4.1) in cases when the required solution is known to be piecewise monotone or convex, while

```
0001            SUBROUTINE PTISR5(A,Z,N,M,IC,L,ISL)
        C * transformation of the operator matrix A(M,N) and the
        C * initial approximation when solving the problem
        C * for piecewise monotone and convex functions
0002            DIMENSION A(M,N), Z(N), ISL(N)
0003            REAL*8 A, Z, R
0004            INTEGER N, M, L, I, J, ISL, J0, IC
0005            J0=1
0006            R=1.
0007            DO 1 J=1,N
0008            IF (J0.EQ.L+1) GOTO 2
0009            IF (J.NE.ISL(J0)) GOTO 2
0010            J0=J0+1
0011            R=-R
0012    2       CONTINUE
0013            Z(J)=R*Z(J)
0014            DO 3 I=1,M
0015            A(I,J)=R*A(I,J)
0016    3       CONTINUE
0017    1       CONTINUE
0018            RETURN
0019            END
```

FIGURE 4.9

its extrema or inflection points are known. In this case the problem of minimizing the discrepancy functional on the corresponding set may also lead to the problem of minimizing the quadratic function $\varphi(z)$ on the set Π^+.

For example, suppose that we know that the unknown function is piecewise monotone and has extrema located at the points s_i^e ($i = 1, \ldots, l$), or is piecewise convex with l inflection points s_i^i ($i = 1, \ldots, l$). We will assume that $a < s_1^i < \cdots < s_l^i < b$; similarly, $a < s_1^e < \cdots < s_l^e < b$. Without loss of generality we may assume that the unknown function is nonincreasing on the interval (a, s_1^e), or, for a piecewise convex function $z(s)$, that it is upper convex on (a, s_1^i).

To solve (4.1) on the set of piecewise monotone or piecewise convex functions, we have to include the following parameters as formal parameters into the program PTISR:

L is an input parameter; it is the number of extrema or inflection points of the unknown function.

ISL is an integer input array; it contains the indexes of extrema or inflection points of the unknown function on the uniform grid $\{s_i\}_{i=1}^n$, $s_1 = a$, $s_n = b$. The indexes in ISL must strictly increase.

The meaning and values of all other parameters is preserved. For IC $= 1$ we look for a piecewise monotone, and for IC $= 3$ for a piecewise convex, solution. We have

```
0001          SUBROUTINE PTISR6(Z,N,IC,L,ISL)
       C * inverse transformation of the solution
       C * for piecewise monotone (convex) functions
0002          DIMENSION Z(N), ISL(N)
0003          REAL*8 Z, R
0004          INTEGER N, L, I, J, ISL, J0, IC
0005          J0=1
0006          R=1.
0007          DO 1 J=1,N
0008          IF (J0.EQ.L+1) GOTO 2
0009          IF (J.NE.ISL(J0)) GOTO 2
0010          J0=J0+1
0011          R=-R
0012     2    CONTINUE
0013          Z(J)=R*Z(J)
0014     1    CONTINUE
0015          RETURN
0016          END
```

FIGURE 4.10

to make the following changes to PTISR:

1. After calling PTISR2 (line 19) we have to call PTISR5:
   CALL PTISR5(R(NA), Z, N, M, IC, L, ISL).
   See Figure 4.9 for the listing of PTISR5.
2. At the end of the program, before calling PTISR3 (line 25) we have to call PTISR6:
   CALL PTISR6(Z, N, M, IC, L, ISL).
   See Figure 4.10 for the listing of PTISR5.
3. Just after calling calling PTISR1 (line 22) we have to make the following assignments:

```
N1=2
IF(IC.EQ.1)N1=1
N2=N-1
```

4. In PTISR1 the lines 24 and 59 have to be changed into:

```
DO  4  N1,N2
```

and

```
DO  10 N1,N2
```

respectively.

5. At the beginning of PTISR and PTISR1 we have to insert a COMMON-block:

COMMON N1, N2

As test computation we propose to solve (4.1) with kernel (4.2) and interval boundaries $a = 0$, $b = 1$, $c = -2$, $d = 2$. The exact solution is taken to be $\overline{z}(s) = 4s(1 - s)$. We will use the a priori information that this solution is convex. As righthand side of the equation we take the values computed in accordance with (4.3). In this case we will put the accuracy δ^2 of specifying the righthand side equal to zero.

Applying PTISR to solve this problem makes it possible to find an approximate solution $z(s)$ whose numerical values have been given in Figure 4.11 (the program call is also given there). The value of the discrepancy $\Phi(z) = \|Az - u_\delta\|^2$ on the approximate solution found is $9.46 \cdot 10^{-19}$, which corresponds to a relative error of approximately $5 \cdot 10^{-9}\%$ (in relation to the maximum of the righthand side). All in all we needed 90 iterations to find the solution. As initial approximation we took $z^{(0)} \equiv 0$.

```
00001          IMPLICIT REAL*8 (A-H,O-Z)
00002          IMPLICIT INTEGER*4(I-N)
00003          DIMENSION U0(41),Z(41),U(41,260),Z0(41)
00004          EXTERNAL AK
00005          X1=0.
00006          X2=1.
00007          Y1=-2.
00008          Y2=2.
00009          N=41
00010          M=41
00011          IMAX=400
00012          IC=3
00013          DL=0.
00014          DO 1 J=1,N
00015          X=X1+(X2-X1)/(N-1.)*(J-1.)
00016        1 Z0(J)=4.*X*(1.-X)
00017          CALL PTICR0(AK,U,X1,X2,Y1,Y2,N,M)
00018          CALL PTICR3(U,Z0,U0,N,M)
00019          DO 4 J=1,N
00020          Z(J)=0.
00021        4 CONTINUE
00022          CALL PTISR(AK,U0,X1,X2,Y1,Y2,N,M,Z,AN,
               *ITER,DL,IMAX,IC,U,41*260,IERR)
00023          PRINT 5,Z0,Z,DL,AN,ITER,IERR
00024          STOP
00025        5 FORMAT(15X,'Exact Solution:'/'.'/
               *                    8(5F11.6/),F11.6/'.'/
               *15X,'Approximate solution:'/'.'/
               *                    8(5F11.6/),F11.6/'.'/
               *10X,'Square of the error of the righthand side
               *                         D14.6/
               *10X,'Discrepancy                   :',
               *                         D14.6/
               *10X,'Number of iterations          :',
               *                         I14/
               *10X,'Return code (IERR)            :',
               *                         I14)
00026          END
```

```
00001          FUNCTION AK(X,Y)
00002          IMPLICIT REAL*8(A-H,O-Z)
00003          AK=1./(1.+100.*(X-Y)**2)
00004          RETURN
00005          END
```

Exact solution:
```
    .000000    .097500    .190000    .277500    .360000
    .437500    .510000    .577500    .640000    .697500
    .750000    .797500    .840000    .877500    .910000
    .937500    .960000    .977500    .990000    .997500
   1.000000    .997500    .990000    .977500    .960000
    .937500    .910000    .877500    .840000    .797500
    .750000    .697500    .640000    .577500    .510000
    .437500    .360000    .277500    .190000    .097500
    .000000
```

Approximate solution:
```
    .000000    .097502    .189998    .277498    .360000
    .437501    .510001    .577501    .639999    .697499
    .749999    .797500    .840000    .877500    .910000
    .937500    .960000    .977500    .990000    .997500
   1.000000    .997500    .990000    .977500    .960000
    .937500    .910000    .877500    .840000    .797500
    .749999    .697499    .639999    .577501    .510001
    .437501    .360000    .277498    .189998    .097502
    .000000
```

Square of the error of the righthand side: .000000D+00
Discrepancy   :   .946349D-18
Number of iterations   :   90
Return code (IERR)   :   0

FIGURE 4.11

Appendix: Program listings

I. Program for solving Fredholm integral equations of the first kind, using Tikhonov's method with transformation of the Euler equation to tridiagonal form

```
0001            SUBROUTINE PTIMR(AK,U0,A,B,C,D,N,M,Z,
               *AN2,DL,H,C1,IMAX,ALFA,U,NU,IERR,
               *AMU,C2,ALP,EPSF)
         C * program to solve integral equations of the first kind
         C * uses smoothing functional minimization
         C * uses generalized discrepancy principle to get
         C *                      regularization parameter
         C * uses transformation to bidiagonal system
0002            IMPLICIT REAL*8(A-H,O-Z)
0003            IMPLICIT INTEGER*4(I-N)
0004            EXTERNAL AK
0005            EXTERNAL PTICRO,PTIMRC,PTIMRS,PTIMRD,
               *PTIMRP,PTIMRR,PTIMRQ,
               *PTIMRN,PTIMRK,PTICR6,PTIMRO,PTIMRA,AUMINM
0006            DIMENSION U0(M),Z(N),U(NU)
0007            ICONT=0
         C * ICONT - start/continue mode flag
         C *   ICONT=0 - start mode
         C *   ICONT=1 - continue mode
0008    110 CONTINUE
0009            DU=DSQRT(DL)
0010            DH=DSQRT(H)
         C * work array mapping
         C * name  length  what
         C * AK(D)  N*M  system matrix or/and D=A*INV(S)
         C *             matrix of reflection vectors
         C * C       N  stabilizer diagonal or/and
```

163

```
      C *                   subdiagonal of P'*P
      C *    B       N      supdiagonal of the stabilizer
      C *                   or/and supdiagonal of P'*P
      C *    S1      N      diagonal S, where C=S'*S
      C *    S2      N      supdiagonal S
      C *    P1      N      diagonal of the P-matrix
      C *    P2      N      supdiagonal of the P-matrix
      C *    U1      N      U1=R*D'*U0
      C *    U2      M      U2=Q'*U0,
      C *                   work array for PTIMR0
      C *    A       N      diagonal of P'*P
      C *    G      M(N)    work array,
      C *                   diagonal of P'*P+ALFA*E
      C *    AL      N      sweep coefficients
      C *    BT      N      sweep coefficients
      C *    initializing array
0011         NAK=1
0012         NC=NAK+N*M
0013         NB=NC+N
0014         NS1=NB+N
0015         NS2=NS1+N
0016         NP1=NS2+N
0017         NP2=NP1+N
0018         NU1=NP2+N
0019         NU2=NU1+N
0020         NA=NU2+M
0021         NG=NA+N
0022         NAL=NG+M
0023         NBT=NAL+N
0024         NOM=NBT+N
0025         IF(NOM-1.GT.NU) GOTO 64
      C * K1,K2 - iteration counters
0026         K1=0
0027         K2=0
0028         KJ=0
0029         AMU=0D0
0030         FUNC=0D0
      C * R0 - derivative weight in the norm
0031         R0=1.
0032         HX=(B-A)/(N-1.)
0033         HY=(D-C)/(M-1.)
0034         DD=HX/HY
      C * EPS - accuracy of the solution of the discrepancy equation
0035         EPS=(C1-1.)*DL
```

```
0036            IF(ICONT.EQ.1) GOTO 111
      C * getting matrix of the operator
0037            CALL PTICRO(AK,U(NAK),A,B,C,D,N,M)
      C * getting matrix of the stabilizer
0038            CALL PTIMRC(U(NC),U(NB),N,RO/HX**2)
      C * transforming the stabilizer matrix to the form S'*S
0039            CALL PTIMRS(U(NC),U(NB),U(NS1),U(NS2),N)
      C * calculating D=A*INV(S)
0040            CALL PTIMRD(U(NS1),U(NS2),U(NAK),N,M)
      C * applying QPR-algorithm to the matrix AK
0041            CALL PTIMRO(U(NAK),U(NP1),U(NP2),U(NG),N,M)
      C * getting tridiagonal matrix P'*P
0042            CALL PTIMRP(U(NP1),U(NP2),U(NC),U(NB),U(NA),N,M)
      C * getting U2=Q'*U0 to compute the discrepancy
0043            CALL PTIMRQ(U0,U(NAK),U(NU2),N,M,U(NG))
      C * getting right-hand side of the system
      C * U1=R*D'*U0=P'*U2
0044            CALL PTIMRR(U(NU1),U(NU2),U(NP1),U(NP2),N,M)
0045       111  CONTINUE
0046            NAN=NA+N-1
      C * calculating incompatibility measure
0047            AN1=0.0
0048            IF(N.EQ.M)GOTO 13
0049            CALL PTICR6(U(NU2+N),U(NU2+N),M-N,AN1)
0050            AN1=AN1*HY
0051       13   CONTINUE
0052            IF(C2.LT.1.)  GOTO 71
0053            CALL AUMINM(U(NA),ALP,IMAX,U(NC),
               *U(NB),U(NG),U(NU1),U(NAL),
               *U(NBT),Z,N,AN1,AN2,M,DU,DH,HX,HY,AN4,
               *U(NP1),U(NP2),U(NU2),AMU,C2,KJ,EPSF)
0054            GOTO (70,71,69),KJ
0055       71   CONTINUE
      C * starting regularization parameter loop
0056            DO 22 I=NA,NAN
0057       22   U(I+N)=U(I)+ALFA*DD
      C * solving system by sweep method
0058            CALL PTIMRN(U(NC),U(NB),U(NG),U(NU1),U(NAL),U(NBT),Z,N)
      C * calculating generalized discrepancy
0059            CALL PTIMRA(AN1,AN2,Z,N,M,DU,DH,HX,HY,
               *AN4,U(NP1),U(NP2),U(NU2),FUNC,AMU,C2)
0060            IF(C1.LE.1)GOTO 100
0061            IF(ALFA.EQ.0.)GOTO 68
0062            IF(AN4.GE.EPS) GOTO 11
```

```
       C * if discrepancy < EPS then multiply ALPHA by 2
       C * until it becomes > EPS
0063         ALFA=2.*ALFA
0064         K1=K1+1
0065         IF(K1.EQ.IMAX) GOTO 65
0066         GOTO 13
       C * setting start points for chord method
0067   11    F0=AN4
0068         X0=1./ALFA
0069         ALFA=ALFA*2.
0070         X=1./ALFA
0071         DO 5 I=NA,NAN
0072    5    U(I+N)=U(I)+ALFA*DD
       C * sweep method
0073         CALL PTIMRN(U(NC),U(NB),U(NG),U(NU1),U(NAL),U(NBT),Z,N)
       C * calculating discrepancy AN2
       C * and generalized discrepancy AN4
0074         CALL PTIMRA(AN1,AN2,Z,N,M,DU,DH,HX,HY,
             *AN4,U(NP1),U(NP2),U(NU2),FUNC,AMU,C2)
0075   14    CONTINUE
       C * if discrepancy < EPS then exit
0076         IF(DABS(AN4).LE.EPS) GOTO 100
0077    1    CONTINUE
       C * if discrepancy < -EPS then go to modified chord method
0078         IF(AN4.LT.-EPS) GOTO 2
0079         IF(ALFA.EQ.0.)GOTO 68
0080         K2=K2+1
0081         IF(K2.EQ.IMAX)GO TO 66
       C * chord method formulas
0082         Y=X0-F0/(AN4-F0)*(X-X0)
0083         X0=X
0084         X=Y
0085         F0=AN4
0086         ALFA=1./X
0087         DO 7 I=NA,NAN
0088    7    U(I+N)=U(I)+ALFA*DD
       C * sweep method
0089         CALL PTIMRN(U(NC),U(NB),U(NG),U(NU1),
             *U(NAL),U(NBT),Z,N)
       C * calculating discrepancy AN2
       C * and generalized discrepancy AN4
0090         CALL PTIMRA(AN1,AN2,Z,N,M,DU,DH,HX,HY,
             *AN4,U(NP1),U(NP2),U(NU2),FUNC,AMU,C2)
0091         GOTO 14
```

```
0092    2    CONTINUE
        C * if generalized discrepancy is negative
        C * starting modified chord method
0093         F=AN4
0094   23    Y=X0+F*(X-X0)/(F-F0)
0095         ALFA=1./Y
0096         DO 15 I=NA,NAN
0097   15    U(I+N)=U(I)+ALFA*DD
        C * sweep method
0098         CALL PTIMRN(U(NC),U(NB),U(NG),U(NU1),U(NAL),U(NBT),Z,N)
        C * calculating discrepancy AN2
        C * and generalized discrepancy AN4
0099         CALL PTIMRA(AN1,AN2,Z,N,M,DU,DH,HX,HY,
            *AN4,U(NP1),U(NP2),U(NU2),FUNC,AMU,C2)
        C * if discrepancy > -EPS but < EPS then exit
0100         IF(DABS(AN4).LE.EPS) GOTO 102
0101         K2=K2+1
0102         IF(K2.GE.IMAX) GOTO 67
0103         IF(ALFA.EQ.0.)GOTO 68
0104         IF(AN4.LE.-EPS) GOTO 37
0105         X0=Y
0106         F0=AN4
0107         GOTO 23
0108   37    CONTINUE
        C * changing interval
0109         X=Y
0110         F=AN4
0111         GOTO 23
0112         ENTRY PTIMRE
        C * entry to continue mode of calculations
0113         ICONT=1
0114         GOTO 110
0115   64    CONTINUE
        C * work array is too short
0116         IERR=64
0117         GOTO 101
0118   65    CONTINUE
        C * initial regularization parameter is too small
0119         IERR=65
0120         GOTO 101
0121   66    CONTINUE
        C * IMAX iterations were made
0122         IERR=66
0123         GOTO 101
```

```
0124    67    CONTINUE
      C * IMAX iterations by modified chord method were made
0125          IERR=67
0126          GOTO 101
0127    68    CONTINUE
      C * ALFA=0 is set or found
0128          IERR=68
0129          GOTO 101
0130    69    CONTINUE
      C * cannot localize minimum in IMAX iterations
0131          IERR=69
0132          GOTO 101
0133    70    CONTINUE
      C * cannot find AMU with specified accuracy
0134          IERR=70
0135          GOTO 101
0136   102    CONTINUE
      C * modified chord method was used
0137          IERR=1
0138          GOTO 101
0139   100    CONTINUE
      C * solution was found by the base chord method
0140          IERR=0
0141   101    CONTINUE
      C * inverse transformation to the old variables
0142          CALL PTIMRK(Z,U(NS1),U(NS2),U(NAK),N,M,
             *Z,U(NG))
0143          RETURN
0144          END


0001          SUBROUTINE PTIMRC(C,B,N,RO)
      C * gets stabilizer
      C * C - diagonal, B - supdiagonal
0002          IMPLICIT REAL*8(A-H,O-Z)
0003          IMPLICIT INTEGER*4(I-N)
0004          DIMENSION C(N),B(N)
0005          DO 1 I=1,N
0006          C(I)=1.+2.*RO
0007          B(I)=-RO
0008          IF(I.EQ.1.OR.I.EQ.N) C(I)=1.+RO
0009     1    CONTINUE
0010          RETURN
0011          END
```

```
0001          SUBROUTINE PTIMRS(C,B,S1,S2,N)
      C * transforms stabilizer to the form S'*S,
      C * S1 - diagonal, S2 - supdiagonal
0002          IMPLICIT REAL*8(A-H,O-Z)
0003          IMPLICIT INTEGER*4(I-N)
0004          DIMENSION C(N),B(N),S1(N),S2(N)
0005          S1(1)=DSQRT(C(1))
0006          S2(1)=B(1)/S1(1)
0007          DO 1 I=2,N
0008          S1(I)=DSQRT(C(I)-(S2(I-1))**2)
0009    1     S2(I)=B(I)/S1(I)
0010          RETURN
0011          END


0001          SUBROUTINE PTIMRD(S1,S2,AA,N,M)
      C * forms matrix D=A*INV(S)
      C * S1,S2 - diagonal and supdiagonal of S
0002          IMPLICIT REAL*8(A-H,O-Z)
0003          IMPLICIT INTEGER*4(I-N)
0004          DIMENSION S1(N),S2(N),AA(M,N)
0005          DO 1 J=1,M
0006    1     AA(J,1)=AA(J,1)/S1(1)
0007          DO 2 I=2,N
0008          DO 3 J=1,M
0009          AA(J,I)=(AA(J,I)-S2(I-1)*AA(J,I-1))/S1(I)
0010    3     CONTINUE
0011    2     CONTINUE
0012          RETURN
0013          END


0001          SUBROUTINE PTIMRO(D,P1,P2,A,N,M)
      C * QPR - algorithm for matrix transformation
      C *                         D(M,N)=Q*P*R
      C * P1,P2 - diagonal and supdiagonal of the matrix P
      C * matrix D(M,N) will contain:
      C *   in columns below diagonal -
      C *              left reflection vectors (Q)
      C *   in rows above diagonal -
      C *              right reflection vectors (R)
0002          IMPLICIT REAL*8(A-H,O-Z)
0003          IMPLICIT INTEGER*4(I-N)
0004          EXTERNAL PTIMR1
```

```
0005            DIMENSION D(M,N),P1(N),P2(N),A(M)
0006            DO 6 K=1,N
     C * getting K-th column
0007            DO 11 J=K,M
0008      11    A(J)=D(J,K)
     C * forming reflection vector
0009            CALL PTIMR1(A(K),M-K+1,B)
0010            A(K)=A(K)-B
0011            CALL PTIMR1(A(K),M-K+1,T)
0012            IF(T.EQ.0.)GOTO 12
0013            DO 13 J=K,M
0014      13    A(J)=A(J)/T
0015            GOTO 14
0016      12    CONTINUE
0017            DO 15 J=K,M
0018      15    A(J)=0.
0019      14    CONTINUE
     C * A(K) - A(M) - reflection vector
     C * writing reflection vector to the matrix D
0020            DO 16 J=K,M
0021      16    D(J,K)=A(J)
     C * getting diagonal element - P1
0022            P1(K)=B
     C * reflect columns from K+1 to N
0023            K1=K+1
0024            IF(K1.GT.N)GOTO 20
0025            DO 17 I=K1,N
0026            S=0.
0027            DO 18 J=K,M
0028      18    S=S+D(J,I)*A(J)
0029            DO 19 J=K,M
0030      19    D(J,I)=D(J,I)-2.*S*A(J)
0031      17    CONTINUE
0032      20    CONTINUE
0033            IF(K1.LT.N)GOTO 60
0034            IF(K1.EQ.N)P2(K)=D(K,K1)
0035            GOTO 6
0036      60    CONTINUE
     C * getting K-th row of the matrix D
0037            DO 21 I=K1,N
0038      21    A(I)=D(K,I)
     C * form reflection vector
0039            CALL PTIMR1(A(K1),N-K1+1,B)
0040            A(K1)=A(K1)-B
```

```
0041            CALL PTIMR1(A(K1),N-K1+1,T)
0042            IF(T.EQ.0.)GOTO 22
0043            DO 23 I=K1,N
0044     23     A(I)=A(I)/T
0045            GOTO 24
0046     22     CONTINUE
0047            DO 25 I=K1,N
0048     25     A(I)=0.
0049     24     CONTINUE
         C * B A(K1) - A(N) - reflection vector
         C * writing reflection vector to K-th row of the matrix
0050            DO 26 I=K1,N
0051     26     D(K,I)=A(I)
         C * getting subdiagonal element P2
0052            P2(K)=B
         C * reflecting rows from K+1 to M
0053            DO 27 J=K1,M
0054            S=0.
0055            DO 28 I=K1,N
0056     28     S=S+D(J,I)*A(I)
0057            DO 29 I=K1,N
0058     29     D(J,I)=D(J,I)-2.*S*A(I)
0059     27     CONTINUE
0060     6      CONTINUE
0061            RETURN
0062            END


0001            SUBROUTINE PTIMR1(A,N,S)
         C * gets norm S of the vector A of length N
0002            IMPLICIT REAL*8(A-H,O-Z)
0003            IMPLICIT INTEGER*4(I-N)
0004            DIMENSION A(N)
0005            R=0.
0006            DO 1 I=1,N
0007            R1=DABS(A(I))
0008            IF(R1.GE.R)R=R1
0009     1      CONTINUE
0010            S=0.
0011            IF(R.EQ.0.)GOTO 999
0012            R1=0.
0013            DO 2 I=1,N
0014            R1=R1+(A(I)/R)**2
0015     2      CONTINUE
```

```
0016            S=DSQRT(R1)*R
0017     999    CONTINUE
0018            RETURN
0019            END


0001            SUBROUTINE PTIMRP(P1,P2,A,B,C,N,M)
         C * gets matrix P'*P
         C * C - diagonal, A - subdiagonal,
         C * B - supdiagonal of the result
0002            IMPLICIT REAL*8(A-H,O-Z)
0003            IMPLICIT INTEGER*4(I-N)
0004            DIMENSION P1(N),P2(N),A(N),B(N),C(N)
0005            C(1)=P1(1)**2
0006            B(1)=P1(1)*P2(1)
0007            A(2)=B(1)
0008            C(N)=P1(N)**2+P2(N-1)**2
0009            B(N)=0.
0010            A(1)=0.
0011            N1=N-1
0012            IF(2.GT.N1)GOTO 2
0013            DO 1 I=2,N1
0014            C(I)=P1(I)**2+P2(I-1)**2
0015            B(I)=P1(I)*P2(I)
0016            A(I+1)=B(I)
0017     1      CONTINUE
0018     2      CONTINUE
0019            RETURN
0020            END


0001            SUBROUTINE PTIMRQ(U0,D,U2,N,M,A)
         C * gets vector U2=Q'*U0 for computing the discrepancy
0002            IMPLICIT REAL*8(A-H,O-Z)
0003            IMPLICIT INTEGER*4(I-N)
0004            EXTERNAL PTICR6,PTICR1
0005            DIMENSION U0(M),D(M,N),U2(M),A(M)
         C * moving righthand side U0
0006            DO 1 J=1,M
0007     1      U2(J)=U0(J)
         C * reflecting vector U2
0008            DO 3 K=1,N
0009            DO 4 J=K,M
0010     4      A(J)=D(J,K)
```

```
      C * in  A - reflection vector
      C * reflecting vector U2
0011        CALL PTICR6(A(K),U2(K),M-K+1,S)
0012        CALL PTICR1(U2(K),A(K),U2(K),M-K+1,-2.*S)
0013    3   CONTINUE
0014        RETURN
0015        END


0001        SUBROUTINE PTIMRR(U1,U2,P1,P2,N,M)
      C * gets righthand side of equation
      C * U1=P'*U2 for sweep method
0002        IMPLICIT REAL*8(A-H,O-Z)
0003        IMPLICIT INTEGER*4(I-N)
0004        DIMENSION U2(M),P1(N),P2(N),U1(N)
0005        U1(1)=P1(1)*U2(1)
0006        DO 1 I=2,N
0007        U1(I)=U2(I-1)*P2(I-1)+U2(I)*P1(I)
0008    1   CONTINUE
0009        RETURN
0010        END


0001        SUBROUTINE PTIMRN(A,B,C,F,AL,BT,X,N)
      C * finds solution of the system
      C * A(I)*X(I-1)+C(I)*X(I)+B(I)*X(I+1)=F(I)
      C * by sweep method
0002        IMPLICIT REAL*8(A-H,O-Z)
0003        IMPLICIT INTEGER*4(I-N)
0004        DIMENSION A(N),B(N),C(N),F(N),AL(N),BT(N)
0005        DIMENSION X(N)
0006        AL(1)=0.
0007        BT(1)=0.
0008        DO 1 I=2,N
0009        Y=-(C(I-1)+A(I-1)*AL(I-1))
0010        AL(I)=B(I-1)/Y
0011        BT(I)=(A(I-1)*BT(I-1)-F(I-1))/Y
0012    1   CONTINUE
      C * sweep coefficients are found
0013        X(N)=(A(N)*BT(N)-F(N))/(-C(N)-A(N)*AL(N))
0014        DO 2 I=2,N
0015        J=N-I+2
0016        X(J-1)=AL(J)*X(J)+BT(J)
0017    2   CONTINUE
```

```
0018        RETURN
0019        END


0001        SUBROUTINE PTIMRA(AN1,AN2,Z,N,M,DL,
           *H,HX,HY,AN4,P1,P2,U2,FUNC,AMU,C2)
C * gets discrepancy AN2
C * and generalized discrepancy AN4
C * AN1 - measure of incompatibility of the system
C * DL,H - errors (not squared ones)
C *         in righthand side and operator
0002        IMPLICIT REAL*8(A-H,O-Z)
0003        IMPLICIT INTEGER*4(I-N)
0004        EXTERNAL PTICR6
0005        DIMENSION P1(N),P2(N),Z(N),U2(M)
C * calculating discrepancy
0006        SUM=0.
0007        DO 46 I=2,N
0008     46 SUM=SUM+(P1(I-1)*Z(I-1)+
           +P2(I-1)*Z(I)-U2(I-1))**2
0009        SUM=SUM+(P1(N)*Z(N)-U2(N))**2
0010        IF (C2.GE.1.)  GOTO 6
0011        AN2=SUM*HY+AN1
C * calculating generalized discrepancy
0012        AN4=AN2-DL**2-AN1
0013        IF(H.EQ.0.)GOTO 5
C * if h≠0 then calculating norm of solution
0014        CALL PTICR6(Z,Z,N,S)
0015        S=DSQRT(S*HX)
C * S - W21-norm of solution
0016        AN4=AN2-(DL+H*S)**2-AN1
0017      5 CONTINUE
0018        GOTO 7
0019      6 CONTINUE
C * using incompatibility measure
0020        AN2=SUM*HY
0021        AN3=DSQRT(AN2)
0022        CALL PTICR6(Z,Z,N,S)
0023        S=DSQRT(S*HX)
0024        FUNC=AN3+H*S
0025        AN4=AN3-H*S-DL-AMU
0026      7 CONTINUE
0027        RETURN
0028        END
```

```
0001          SUBROUTINE PTIMRK(X,S1,S2,D,N,M,Z,A)
      C * inverse transformation
      C *           Z=INV(S)*R'*X
0002          IMPLICIT REAL*8(A-H,O-Z)
0003          IMPLICIT INTEGER*4(I-N)
0004          EXTERNAL PTICR6,PTICR1
0005          DIMENSION X(N),S1(N),S2(N),D(M,N),
             *Z(N),A(N)
0006          IF(3.GT.N)GOTO 3
0007          DO 1 II=3,N
0008          K=N-II+1
0009          K1=K+1
0010          DO 2 I=K1,N
0011     2    A(I)=D(K,I)
      C * in  A - reflection vector
      C * reflecting vector X
0012          CALL PTICR6(A(K1),X(K1),N-K1+1,S)
0013          CALL PTICR1(X(K1),A(K1),X(K1),N-K1+1,-2.*S)
0014     1    CONTINUE
0015     3    CONTINUE
      C * multiplying by matrix inverse to S (solving the system)
0016          N1=N-1
0017          Z(N)=X(N)/S1(N)
0018          DO 5 I=1,N1
0019          J=N-I
0020          Z(J)=(X(J)-S2(J)*Z(J+1))/S1(J)
0021     5    CONTINUE
0022          RETURN
0023          END


0001          SUBROUTINE AUMINM(ZV,ALP,KMAX,A,B,C,F,
             *AL,BT,X,N,AN1,AN2,M,DL,H,HX,HY,
             *AN4,P1,P2,U2,AMU,C2,K2,EPSF)
0002          IMPLICIT REAL*8(A-H,O-Z)
0003          IMPLICIT INTEGER*4(I-N)
0004          DIMENSION ZV(N),A(N),B(N),C(N),F(N),AL(N)
0005          DIMENSION BT(N),X(N),P1(N),P2(N),U2(M)
0006          EXTERNAL PTIMRN,PTIMRA
0007          LI=0
0008          DD=HX/HY
0009          K3=0
0010          K2=0
0011          X1=0D0
```

```
0012      1   CONTINUE
0013          DO 13 I=1,N
0014     13   C(I)=ZV(I)+ALP*DD
0015          CALL PTIMRN(A,B,C,F,AL,BT,X,N)
0016          CALL PTIMRA(AN1,AN2,X,N,M,DL,H,HX,HY,
             *AN4,P1,P2,U2,FUNC,AMU,C2)
0017          GOTO (2,5,6),LI
0018          F2=FUNC
0019          X2=ALP
0020          ALP=ALP*2D0
0021          LI=1
0022          GOTO 1
0023      2   F3=FUNC
0024          X3=ALP
0025          IF(F3-F2) 3,3,4
0026      3   F1=F2
0027          X1=X2
0028          F2=F3
0029          X2=X3
0030          ALP=ALP*2D0
0031          K3=K3+1
0032          IF(K3.GE.KMAX) GOTO 15
0033          GOTO 1
0034      4   X4=X3
0035          F4=F3
0036          AU=(DSQRT(5D0)-1D0)/2D0
0037     10   X3=X1+AU*(X4-X1)
0038          X2=X4-AU*(X4-X1)
0039          ALP=X2
0040          LI=2
0041          GOTO 1
0042      5   F2=FUNC
0043          ALP=X3
0044          LI=3
0045          GOTO 1
0046      6   F3=FUNC
0047          IF(DABS(X3-X2).LE.1D0/C2) GOTO 12
0048          IF(X1.EQ.0D0) GOTO 322
0049          IF(DABS(DMIN1(F2,F3)-DMAX1(F1,F4)).LE.EPSF) GOTO 12
0050          GOTO 321
0051    322   IF(F4-DMIN1(F2,F3).LE.EPSF) GOTO 12
0052    321   K2=K2+1
0053          IF(K2.GE.KMAX) GOTO 14
0054          IF(F3-F2) 7,8,9
```

```
0055      7   X1=X2
0056          F1=F2
0057          GOTO 10
0058      9   X4=X3
0059          F4=F3
0060          GOTO 10
0061      8   X4=(X4+X3)/2D0
0062          GOTO 10
0063     14   K2=1
0064     12   ALP=(X3+X2)/2D0
0065          AMU=(F3+F2)/2D0+DL
0066          K2=2
0067          GOTO 16
0068     15   K2=3
0069     16   RETURN
0070          END
```

II. Program for solving Fredholm integral equations of the first kind by Tikhonov's method, using the conjugate gradient method

```
0001          SUBROUTINE PTIZR(AK,U0,A,B,C,D,N,M,Z,
              *AN2,DL,H,C1,IMAX,ALFA,U,NU,IERR)
        C * Program to solve integral equations
        C * of the first kind by Tikhonov's method using the
        C * generalized discrepancy principle.
        C * The Tikhonov functional is minimized by
        C * the method of conjugate gradients
0002          IMPLICIT REAL*8(A-H,O-Z)
0003          IMPLICIT INTEGER*4(I-N)
0004          DIMENSION U0(M),U(NU),Z(N)
0005          EXTERNAL AK
0006          EXTERNAL PTICRO,PTIZR1,PTIZRA
        C *   work array mapping
        C *   name length   what
        C *   A     N*M      operator matrix
        C *   GRD   N        gradient in B PTIZR1
        C *   S     N        descent direction in PTIZR1
        C *   P     M        work array for PTIZR1
0007          ICONT=0
        C * ICONT - start/continue mode flag
        C *   ICONT=0 - start mode
        C *   ICONT=1 - continue mode
0008    110   CONTINUE
0009          DU=SQRT(DL)
```

```
0010          DH=SQRT(H)
       C * initializing array
0011          NA=1
0012          NGRD=NA+N*M
0013          NS=NGRD+N
0014          NP=NS+N
0015          NU1=NP+M
0016          IF(NU1-1.GT.NU) GOTO 64
       C * K1,K2 - iterations counters
0017          K1=0
0018          K2=0
0019          N1=N+1
0020          HX=(B-A)/(N-1.)
0021          HY=(D-C)/(M-1.)
0022          DD=HX/HY
       C * RO - derivative weight in W21-norm
       C * divided by square of the grid step
0023          RO=1./HX**2
       C * EPS - accuracy for discrepancy equation
0024          EPS=(C1-1.)*DL
0025          IF(ICONT.EQ.1) GOTO 111
       C * forming matrix of the operator
0026          CALL PTICRO(AK,U(NA),A,B,C,D,N,M)
0027    111   CONTINUE
       C * getting initial ALFA so that
       C * generalized discrepancy will be positive
0028     13   CONTINUE
       C * minimizing functional by conjugate gradient method
0029          CALL PTIZR1(U(NA),Z,U0,N,M,ITER,
              *0D0,0D0,N,AN2,ALFA*DD,RO,Z,
              *U(NGRD),U(NS),U(NP),IER)
       C * calculating generalized discrepancy
0030          CALL PTIZRA(AN2,Z,N,DU,DH,HX,HY,RO,AN4)
0031          IF(C1.LE.1.)GOTO 100
0032          IF(ALFA.EQ.0.)GOTO 68
0033          IF(AN4.GT.EPS) GOTO 11
       C * if < EPS then multiply ALFA by 2
0034          K1=K1+1
0035          IF(K1.EQ.IMAX) GOTO 65
0036          ALFA=2.*ALFA
0037          GOTO 13
       C * setting two initial points for chord method
0038     11   CONTINUE
0039          F0=AN4
```

```
0040          X0=1./ALFA
0041          ALFA=ALFA*2.
0042          X=1./ALFA
       C * conjugate gradient method
0043          CALL PTIZR1(U(NA),Z,U0,N,M,ITER,
              *0D0,0D0,N,AN2,ALFA*DD,RO,Z,
              *U(NGRD),U(NS),U(NP),IER)
       C * calculating generalized discrepancy
0044          CALL PTIZRA(AN2,Z,N,DU,DH,HX,HY,RO,AN4)
0045    14    CONTINUE
       C * if accuracy achieved then exit
0046          IF(DABS(AN4).LE.EPS) GOTO 100
       C * if discrepancy < -EPS then starting modified chord method
0047          IF(AN4.LE.-EPS)GOTO 2
0048          IF(ALFA.EQ.0.)GOTO 68
0049          K2=K2+1
0050          IF(K2.EQ.IMAX)GOTO 66
       C * chord method formulas
0051          Y=X0-F0/(AN4-F0)*(X-X0)
0052          X0=X
0053          X=Y
0054          F0=AN4
0055          ALFA=1./X
       C * conjugate gradient method
0056          CALL PTIZR1(U(NA),Z,U0,N,M,ITER,
              *0D0,0D0,N,AN2,ALFA*DD,RO,Z,
              *U(NGRD),U(NS),U(NP),IER)
       C * calculating generalized discrepancy
0057          CALL PTIZRA(AN2,Z,N,DU,DH,HX,HY,RO,AN4)
0058          GOTO 14
0059     2    CONTINUE
       C * starting modified chord method
0060          F=AN4
0061    23    CONTINUE
0062          Y=X0+F*(X-X0)/(F-F0)
0063          ALFA=1./Y
       C * conjugate gradient method
0064          CALL PTIZR1(U(NA),Z,U0,N,M,ITER,
              *0D0,0D0,N,AN2,ALFA*DD,RO,Z,
              *U(NGRD),U(NS),U(NP),IER)
       C * calculating generalized discrepancy
0065          CALL PTIZRA(AN2,Z,N,DY,DH,HX,HY,RO,AN4)
       C * if accuracy achieved then exit
0066          IF(DABS(AN4).LE.EPS) GOTO 101
```

```
0067            IF(AN4.LE.-EPS) GOTO 37
0068            IF(ALFA.EQ.0.)GOTO 68
0069            K2=K2+1
0070            IF(K2.EQ.IMAX)GOTO 67
0071            XO=Y
0072            FO=AN4
0073            GOTO 23
0074      37    CONTINUE
        C * changing interval
0075            X=Y
0076            F=AN4
0077            GOTO 23
0078            ENTRY PTIZRE
        C * entry to continue calculations
0079            ICONT=1
0080            GOTO 110
0081      64    CONTINUE
        C * work array is too short
0082            IERR=64
0083            GOTO 999
0084      65    CONTINUE
        C * initial regularization parameter is too small
0085            IERR=65
0086            GOTO 999
0087      66    CONTINUE
        C * IMAX iterations of the chord method made
0088            IERR=66
0089            GOTO 999
0090      67    CONTINUE
        C * IMAX iterations by the modified chord method made
0091            IERR=67
0092            GOTO 999
0093      68    CONTINUE
        C * ALFA=0 is specified or found
0094            IERR=68
0095            GOTO 999
0096     100 CONTINUE
        C * solution is found by the chord method
0097            IERR=0
0098            GOTO 999
0099     101    CONTINUE
        C * solution is found by the modified chord method
0100            IERR=1
0101     999    CONTINUE
```

```
0102          RETURN
0103          END


0001          SUBROUTINE PTIZR1(A,Z0,U0,N,M,ITER,DL2,
             *ANGRD,IMAX,AN2,ALF,RO,Z,GRAD,S,U,IERR)
      C * minimizes the Tikhonov functional
      C * by the conjugate gradient method.
      C * AK(M,N) - operator matrix in equation AZ=U
      C * Z0 - start point
      C * U0 - righthand side of the equation AZ=U
      C * DL2 - discrepancy level to stop iterations
      C * ANGRD - level of the norm of the gradient
      C * AN2 - discrepancy achieved
      C * ALF - regularization parameter
      C * RO - differential derivative weight
      C *              in stabilizer
      C * solution - extremal in the array Z
0002          IMPLICIT REAL*8(A-H,O-Z)
0003          IMPLICIT INTEGER*4(I-N)
0004          DIMENSION A(M,N),Z0(N),U0(M),
             *Z(N),GRAD(N),S(N),U(M)
0005          EXTERNAL PTICR1,PTICR4,PTICR8,
             *PTICR6,PTICR3,PTICR5,PTICR0
0006          ITER=0
0007          CALL PTICR1(Z0,Z0,Z,N,ODO)
      C * getting gradient of functional
0008          CALL PTICR3(A,Z,U,N,M)
0009          CALL PTICR4(GRAD,U,U0,A,N,M)
0010          CALL PTICR8(GRAD,Z,N,ALF,RO)
      C * getting conjugate direction S
0011          CALL PTICR1(GRAD,GRAD,S,N,ODO)
      C * getting norm of S
0012          CALL PTICR6(S,S,N,RNORM1)
      C * Attention!  Machine constant
0013          ALM=1.E+18
      C * solving one-dimensional minimization problem
      C * getting descent step
0014     7 CONTINUE
0015          CALL PTICR0(A,Z,GRAD,U,S,ALM,BETA,
             *N,M,ALF,RO,IED)
0016          IF(BETA.EQ.0.)GOTO 14
0017          BETA=-BETA
      C * getting new approximation Z
```

```
0018            CALL PTICR1(Z,S,Z,N,BETA)
      C * getting discrepancy
0019            CALL PTICR3(A,Z,U,N,M)
0020            CALL PTICR5(U,U0,M,AN2)
      C * if accuracy achieved then exit
0021            IF(AN2.LE.DL2)GOTO 10
0022            ITER=ITER+1
      C * getting gradient
0023            CALL PTICR4(GRAD,U,U0,A,N,M)
0024            CALL PTICR8(GRAD,Z,N,ALF,RO)
      C * getting gradient norm
0025            CALL PTICR6(GRAD,GRAD,N,RNORM)
0026            RNM=RNORM/RNORM1
      C * getting conjugate direction
0027            CALL PTICR1(GRAD,S,S,N,RNM)
0028            RNORM1=RNORM
      C * if gradient norm is too small then exit
0029            IF(RNORM1.LE.ANGRD) GOTO 11
0030            IF(ITER.GE.IMAX) GOTO 13
0031            GOTO 7
0032   10       CONTINUE
      C * accuracy achieved
0033            IERR=0
0034            GOTO 999
0035   11       CONTINUE
      C * gradient norm is small
0036            IERR=1
0037            GOTO 999
0038   13       CONTINUE
      C * IMAX iterations made
0039            IERR=2
0040            GOTO 999
0041   14       CONTINUE
      C * minimization step equals zero
0042            IERR=3
0043   999      CONTINUE
0044            RETURN
0045            END


0001            SUBROUTINE PTIZRA(AN2,Z,N,DL,DH,
               *HX,HY,RO,AN4)
      C * calculates generalized discrepancy
0002            IMPLICIT REAL*8(A-H,O-Z)
```

```
0003            IMPLICIT INTEGER*4(I-N)
0004            DIMENSION Z(N)
0005            EXTERNAL PTICR6
      C * norm of the discrepancy
0006            AN2=AN2*HY
0007            AN4=AN2-DL**2
      C * if H=0 the norm of the solution need not be calculated
0008            IF(DH.EQ.0.)GOTO 999
      C * calculating the W21-norm of the solution
0009            CALL PTICR6(Z,Z,N,S)
0010            S=S*HX
      C * S - square of the L2-norm of the solution
0011            S1=0.
0012            DO 1 I=2,N
0013            S1=S1+(Z(I)-Z(I-1))**2
0014      1     CONTINUE
0015            S1=S1*HX
      C * S1 - square of the L2-norm of the derivative
      C *                                  of the solution
0016            S=DSQRT(S+RO*S1)
      C * S - W21-norm of the solution
0017            AN4=AN2-(DL+DH*S)**2
0018      999   CONTINUE
0019            RETURN
0020            END
```

III. Program for solving Fredholm integral equations of the first kind on the set of nonnegative functions, using the regularization method

```
0001            SUBROUTINE PTIPR(AK,UO,A,B,C,D,N,M,Z,
               *IC,AN2,DL,H,C1,ANGRD,IMAX,
               *ALFA,R,NR,IERR)
      C * Program to solve integral equations
      C * of the first kind by the regularization method
      C * with generalized discrepancy principle to get
      C * regularization parameter.
      C * The conjugate gradient method is used
      C * to minimize the smoothing functional
0002            IMPLICIT REAL*8(A-H,O-Z)
0003            IMPLICIT INTEGER*4(I-N)
0004            DIMENSION UO(M),R(NR),Z(N)
0005            EXTERNAL AK
0006            EXTERNAL PTICR0,PTICR1,
               *PTISR1,PTISR2,PTISR3
```

```
       C * Work array mapping
       C *    name      length          what
       C *     A:        N*M            operator matrix
       C *     H:        N              descent direction
       C *     G:        N              gradient
       C *     U:        M              operator value
       C *     U1:       M              work array
       C *     S:        N              work array
       C *
       C *  NR=N*M+3N+2M
0007          ICONT=0
       C * ICONT - start/continue mode flag
       C *    ICONT=0   start mode
       C *    ICONT=1   continue mode
0008   110  CONTINUE
0009          IF(IC.NE.0.AND.IC.NE.1)GOTO 69
       C * initializing arrays
0010          NA=1
0011          NH=N*M+1
0012          NG=NH+N
0013          NU=NG+N
0014          NU1=NU+M
0015          NS=NU1+M
0016          NMAX=NS+N
0017          IF(NMAX-1.GT.NR)GOTO 64
0018          DU=DSQRT(DL)
0019          DH=DSQRT(H)
       C * K1,K2 - iteration numbers
0020          K1=0
0021          K2=0
0022          N1=N+1
0023          HX=(B-A)/(N-1.)
0024          HY=(D-C)/(M-1.)
0025          DD=HX/HY
       C * RO - derivative weight in the norm
       C * divided by square of the grid step
0026          IF(IC.EQ.0)RO=1./HX**2
0027          IF(IC.EQ.1)RO=0.0
       C * EPS - accuracy of the solution of the discrepancy equation
0028          EPS=(C1-1.)*DL
0029          IF(ICONT.EQ.1) GOTO 111
       C * getting operator matrix A
0030          CALL PTICRO(AK,R(NA),A,B,C,D,N,M)
0031   111  CONTINUE
```

```
      C * transition to PI-plus
0032        CALL PTISR2(R(NA),Z,N,M,IC,R(NS))
0033        CALL PTICR1(Z,Z,R(NH),N,ODO)
      C * finding regularization parameter with
      C * positive generalized discrepancy
0034   13   CONTINUE
      C * get minimum of the functional by the
      C *            conjugate gradient method
0035        CALL PTISR1(R(NA),Z,UO,N,M,ITER,
          *ODO,ODO,N,AN2,ALFA*DD,RO,Z,R(NU),
          *R(NU1),R(NH),R(NG),R(NS),IER)
      C * calculating generalized discrepancy
0036        CALL PTIZRA(AN2,Z,N,DU,DH,HX,HY,RO,AN4)
0037        IF(C1.LE.1.)GOTO 100
0038        IF(ALFA.EQ.0.)GOTO 68
0039        IF(AN4.GT.EPS) GOTO 11
      C * multiply ALFA by 2 while discrepancy < EPS
0040        K1=K1+1
0041        IF(K1.EQ.IMAX) GOTO 65
0042        ALFA=2.*ALFA
0043        GOTO 13
      C * setting initial points of the chord method
0044   11   CONTINUE
0045        FO=AN4
0046        X0=1./ALFA
0047        ALFA=ALFA*2.
0048        X=1./ALFA
      C * using conjugate gradient method to minimize functional
0049        CALL PTISR1(R(NA),Z,UO,N,M,ITER,
          *ODO,ODO,N,AN2,ALFA*DD,RO,Z,R(NU),
          *R(NU1),R(NH),R(NG),R(NS),IER)
      C * getting generalized discrepancy
0050        CALL PTIZRA(AN2,Z,N,DU,DH,HX,HY,RO,AN4)
0051   14   CONTINUE
      C * if accuracy achieved then exit
0052        IF(DABS(AN4).LE.EPS) GOTO 100
      C * if discrepancy < 0 then go to modified chord method
0053        IF(AN4.LE.-EPS)GOTO 2
0054        IF(ALFA.EQ.0.)GOTO 68
0055        K2=K2+1
0056        IF(K2.EQ.IMAX)GOTO 66
      C * chord method formulas
0057        Y=X0-FO/(AN4-FO)*(X-X0)
0058        X0=X
```

```
0059            X=Y
0060            F0=AN4
0061            ALFA=1./X
        C * using conjugate gradient method
0062            CALL PTISR1(R(NA),Z,U0,N,M,ITER,
                *OD0,OD0,N,AN2,ALFA*DD,RO,Z,R(NU),
                .*R(NU1),R(NH),R(NG),R(NS),IER)
        C * calculating generalized discrepancy
0063            CALL PTIZRA(AN2,Z,N,DU,DH,HX,HY,RO,AN4)
0064            GOTO 14
0065         2  CONTINUE
        C * starting modified chord method
0066            F=AN4
0067        23  CONTINUE
0068            Y=X0+F*(X-X0)/(F-F0)
0069            ALFA=1./Y
        C * using conjugate gradient method
0070            CALL PTISR1(R(NA),Z,U0,N,M,ITER,
                *OD0,OD0,N,AN2,ALFA*DD,RO,Z,R(NU),
                *R(NU1),R(NH),R(NG),R(NS),IER)
        C * getting generalized discrepancy
0071            CALL PTIZRA(AN2,Z,N,DY,DH,HX,HY,RO,AN4)
        C * if accuracy achieved then exit
0072            IF(DABS(AN4).LE.EPS) GOTO 101
0073            IF(AN4.LE.-EPS) GOTO 37
0074            IF(ALFA.EQ.0.)GOTO 68
0075            K2=K2+1
0076            IF(K2.EQ.IMAX)GOTO 67
0077            X0=Y
0078            F0=AN4
0079            GOTO 23
0080        37  CONTINUE
        C * changing the interval
0081            X=Y
0082            F=AN4
0083            GOTO 23
0084            ENTRY PTIPRE
        C * entry to continue calculations
0085            ICONT=1
0086            GOTO 110
0087        64  CONTINUE
        C * work array is too short
0088            IERR=64
0089            GOTO 9999
```

```
0090   65   CONTINUE
       C * initial regularization parameter is too small
0091        IERR=65
0092        GOTO 999
0093   66   CONTINUE
       C * IMAX iterations of chord method made
0094        IERR=66
0095        GOTO 999
0096   67   CONTINUE
       C * IMAX iterations of modified chord method made
0097        IERR=67
0098        GOTO 999
0099   68   CONTINUE
       C * ALFA=0 is set or found
0100        IERR=68
0101        GOTO 999
0102   69   CONTINUE
       C * incorrect set type is specified
0103        IERR=69
0104        GOTO 9999
0105   100  CONTINUE
       C * solution is found
0106        IERR=0
0107        GOTO 999
0108   101  CONTINUE
       C * solution is found by modified chord method
0109        IERR=1
0110   999  CONTINUE
       C * return from PI-plus to original coordinates
0111        CALL PTICR1(Z,Z,R(NH),N,ODO)
0112        CALL PTISR3(Z,N,IC,R(NS))
0113  9999  CONTINUE
0114        RETURN
0115        END


0001        SUBROUTINE PTIZRA(AN2,Z,N,DL,DH,
            *HX,HY,RO,AN4)
       C * calculates generalized discrepancy
0002        IMPLICIT REAL*8(A-H,O-Z)
0003        IMPLICIT INTEGER*4(I-N)
0004        DIMENSION Z(N)
0005        EXTERNAL PTICR6
       C * norm of the discrepancy
```

```
0006           AN2=AN2*HY
0007           AN4=AN2-DL**2
     C * if H=0 then no need to calculate the norm of the solution
0008           IF(DH.EQ.0.)GOTO 999
     C * calculating the W21-norm of the solution
0009           CALL PTICR6(Z,Z,N,S)
0010           S=S*HX
     C * S - square of the L2-norm of the solution
0011           S1=0.
0012           DO 1 I=2,N
0013           S1=S1+(Z(I)-Z(I-1))**2
0014     1     CONTINUE
0015           S1=S1*HX
     C * S1 - square of the L2-norm of the derivative
     C *                                of the solution
0016           S=DSQRT(S+RO*S1)
     C * S - W21-norm of the solution
0017           AN4=AN2-(DL+DH*S)**2
0018     999   CONTINUE
0019           RETURN
0020           END
```

IV. Program for solving one-dimensional integral equations of convolution type

```
0001           SUBROUTINE PTIKR(AK,U0,A,B,C,D,
               *L1,L2,N,Z,AN,DL,HH,C1,IMAX,ALPHA,
               *U,NU,IERR)
0002           IMPLICIT REAL*8(A-H,O-Z)
0003           IMPLICIT INTEGER*4(I-N)
0004           REAL*8 L1,L2
0005           DIMENSION U(NU),U0(N),Z(N)
0006           EXTERNAL AK
0007           EXTERNAL PTICR1,PTICR2,PTIKR1
     C * work array mapping
     C * name length    what it is
     C *  ARE   N    real part of the FT of the kernel
     C *  AIM   N    imaginary part of the FT of the kernel
     C *  URE   N    real part of the FT of the righthand side
     C *  UIM   N    imaginary part of the FT of the righthand side
     C *  ZIM   N    imaginary part of the solution
     C *  W     N    stabilizer
0008           IP=1
     C * IP - mode:
```

```
      C * at the first call IP=+1
      C * at each subsequent call (up till PTIKRE) IP=-1
0009        EPRO=0.
0010        IF(C1.GT.1.)EPRO=(C1-1.)*DL
      C * EPRO - accuracy for discrepancy equation
      C * if EPRO=0 then the extremal is only calculated with the
      C * fixed regularization parameter ALFA
0011  100   CONTINUE
      C * calculating support of the solution
0012        A=C-.5*(L1+L2)
0013        B=D-.5*(L1+L2)
      C * T - period
0014        T=D-C
      C * initializing arrays
0015        NAR=1
0016        NAI=NAR+N
0017        NW=NAI+N
0018        NZI=NW+N
0019        NUR=NZI+N
0020        NUI=NUR+N
0021        NMAX=NUI+N
      C * check sufficiency of the size of the work array
0022        IF(NMAX-1.GT.NU) GO TO 64
0023        IF(IP.EQ.-1) GO TO 101
      C * setting righthand side
0024        CALL PTICR1(UO,UO,U(NUR),N,ODO)
0025        CALL PTICR2(U(NUI),ODO,N)
0026  101   CONTINUE
      C * solving equation
0027        CALL PTIKR1(AK,U(NAR),U(NAI),Z,
           *U(NZI),U(NUR),U(NUI),U(NW),N,ALPHA,
           *L1,L2,AN,OM,T,DSQRT(DL),DSQRT(HH),
           *IP,EPRO,IQ,IMAX,IERR)
0028        GO TO 999
0029   64   CONTINUE
      C * work array is too short
0030        IERR=64
0031        GO TO 999
      C * the entry PTIKRE is for repeatedly solving
      C * the same equation and righthand side
0032        ENTRY PTIKRE
0033        IP=-1
0034        GO TO 100
0035  999   CONTINUE
```

```
0036          RETURN
0037          END


0001          SUBROUTINE PTIKR1(AK,ARE,AIM,ZRE,ZIM,
             *URE,UIM,W,N,ALP,L1,L2,BETA,RO,A,
             *DEL,HH,IPAR,EPRO,IQ,IMAX,IERR)
0002          IMPLICIT REAL*8(A-H,O-Z)
0003          IMPLICIT INTEGER*4(I-N)
0004          REAL*8 L1,L2
0005          DIMENSION ARE(N),ZRE(N),URE(N),
             *AIM(N),ZIM(N),UIM(N),W(N)
0006          EXTERNAL FTF1C
0007          H=A/N
0008          HA=H/N
      C * checking if it is the first call
0009          IF(IPAR.EQ.-1) GO TO 2
      C * getting kernel of the equation
0010          DO 1 I=1,N
0011          ARG=(I-N/2-1)*H+0.5*(L1+L2)
0012          ARE(I)=AK(ARG)*H
0013          AIM(I)=0.0
0014          IF(ARG.LT.L1.OR.ARG.GT.L2) ARE(I)=0.0
0015          W(I)=(2.0/H*DSIN(3.14159265358D0/
             /N*(I-1)))**2
0016        1 CONTINUE
0017          P=1.0
      C * Fourier transformation of the kernel and righthand side
0018          CALL FTF1C(ARE,AIM,N,1,1,P)
0019          CALL FTF1C(URE,UIM,N,1,1,P)
0020        2 CONTINUE
      C * getting ALP so that RO(ALP) > 0
      C * IQ - iteration number
0021          IQ=0
0022       77 CONTINUE
0023          IQ=IQ+1
      C * calculating the discrepancy
0024          F1=0.0
0025          F2=0.0
0026          DO 44 M=1,N
0027          U2=URE(M)**2+UIM(M)**2
0028          A2=ARE(M)**2+AIM(M)**2
0029          X=1.0+W(M)
0030          IF(A2.EQ.0.0) GO TO 42
```

```
0031          BA=X/(A2+ALP*X)
0032          AB=1.0-ALP*BA
0033          C1=HA*U2*(BA*ALP)**2
0034          C2=HA*U2*AB*BA
0035          F1=F1+C1
0036          F2=F2+C2
0037          GO TO 44
0038      42 F1=F1+HA*U2
0039      44 CONTINUE
      C * calculating the generalized discrepancy
      C * to check if RO(ALP)>0
0040          BETA=F1
0041          RO=F1-(DEL+HH*DSQRT(F2))**2
      C * if EPRO=0 is set then finish
0042          IF(EPRO.EQ.0.0) GO TO 10
      C * if ALP=0.0 is set then finish
0043          IF(ALP.EQ.0.0) GO TO 68
0044          IF(RO.GT.0.0) GO TO 33
      C * go to calculation of the solution if there is no
      C * parameter ALP with RO(ALP) > 0
0045          IF(IQ.GT.IMAX) GO TO 65
0046          ALP=2.0*ALP
0047          GO TO 77
      C * starting Newton method
0048      33 CONTINUE
0049          IQ=0
0050       3 CONTINUE
0051          IQ=IQ+1
      C * calculating the discrepancy
0052          F1=0.0
0053          F2=0.0
0054          F3=0.0
0055          DO 4 M=1,N
0056          U2=URE(M)**2+UIM(M)**2
0057          A2=ARE(M)**2+AIM(M)**2
0058          X=1.0+W(M)
0059          IF(A2.EQ.0.0) GO TO 41
0060          BA=X/(A2+ALP*X)
0061          AB=1.0-ALP*BA
0062          C1=HA*U2*(BA*ALP)**2
0063          C2=HA*U2*AB*BA
0064          C3=2.0*C1*AB
0065          F1=F1+C1
0066          F2=F2+C2
```

```
0067          F3=F3+C3
0068           GO TO 4
0069        41 F1=F1+HA*U2
0070         4 CONTINUE
      C * calculating the discrepancy - BETA,
      C * W21-norm of the solution - ZNOR,
      C * generalized discrepancy - RO
      C * and its derivative - DR
0071           BETA=F1
0072           ZNOR=DSQRT(F2)
0073           RO=BETA-(DEL+HH*ZNOR)**2
0074           IF(ALP.EQ.0.0) GO TO 68
0075           DR=-F3*ALP-(DEL+HH*ZNOR)*HH*F3/ZNOR
      C * calculate solution if accuracy achieved
0076           IF(DABS(RO).LT.EPRO) GO TO 10
      C * starting chord method if generalized discrepancy
      C * is not convex
0077           IF(RO.LT.0.0) GO TO 61
      C * getting new argument by Newton's method
0078           DQ=-RO/DR
0079           A1=ALP
0080           R1=RO
0081           ALP=1.0/(1.0/ALP+DQ)
0082           IF(IQ.GE.IMAX) GO TO 66
0083           GO TO 3
      C * chord method
0084        61 CONTINUE
      C * change interval if RO<0
0085         6 CONTINUE
0086           AO=ALP
0087           RO=RO
0088         7 CONTINUE
0089           IQ=IQ+1
      C * getting new argument by chord method
0090           ALP=AO*A1*(RO-R1)/(AO*RO-A1*R1)
      C * calculating discrepancies
0091         8 CONTINUE
0092           F1=0.0
0093           F2=0.0
0094           DO 9 M=1,N
0095           U2=URE(M)**2+UIM(M)**2
0096           A2=ARE(M)**2+AIM(M)**2
0097           X=1.0+W(M)
0098           IF(A2.EQ.0.0) GO TO 91
```

```
0099          BA=X/(A2+ALP*X)
0100          AB=1.0-ALP*BA
0101          C1=HA*U2*(BA*ALP)**2
0102          C2=HA*U2*AB*BA
0103          F1=F1+C1
0104          F2=F2+C2
0105          GO TO 9
0106       91 F1=F1+HA*U2
0107        9 CONTINUE
C * generalized discrepancy and norm of the solution
0108          BETA=F1
0109          ZNOR=DSQRT(F2)
0110          RO=BETA-(DEL+HH*ZNOR)**2
0111          IF(ALP.EQ.0.0) GO TO 68
C * calculating solution if accuracy achieved
0112          IF(DABS(RO).LT.EPRO) GO TO 11
0113          IF(IQ.EQ.IMAX) GO TO 67
0114          IF(RO.LT.0.0) GO TO 6
C * change interval if RO>0
0115          A1=ALP
0116          R1=RO
0117          GO TO 7
0118       65 CONTINUE
C * RO(ALP) <= 0 for all regularization parameters
0119          IERR=65
0120          GO TO 999
0121       66 CONTINUE
C * IMAX iterations by Newton's method made
0122          IERR=66
0123          GO TO 999
0124       67 CONTINUE
C * IMAX iterations by chord method made
0125          IERR=67
0126          GO TO 999
0127       68 CONTINUE
C * ALP=0.0 is set or found
0128          IERR=68
0129          GO TO 999
0130       11 CONTINUE
C * solution is found by chord method
0131          IERR=1
0132          GO TO 999
0133       10 CONTINUE
C * solution is found by Newton's method
```

```
0134          IERR=0
0135      999 CONTINUE
C * calculating Fourier transform of the solution
0136          SSI=-1.0
0137          DO 12 M=1,N
0138          SSI=-SSI
0139          ZZ=N*(ARE(M)**2+AIM(M)**2+ALP*(1.0+W(M)))
0140          IF(ZZ.NE.0.0) GO TO 111
0141          ZRE(M)=0.0
0142          ZIM(M)=0.0
0143          GO TO 12
0144      111 ZRE(M)=SSI*(ARE(M)*URE(M)+AIM(M)*UIM(M))/ZZ
0145          ZIM(M)=SSI*(ARE(M)*UIM(M)-AIM(M)*URE(M))/ZZ
0146       12 CONTINUE
C * inverse Fourier transform of the solution
0147          P=-1.0
0148          CALL FTF1C(ZRE,ZIM,N,1,1,P)
0149          RETURN
0150          END


0001          SUBROUTINE FTF1C(ARE,AIM,N,IN,K,P)
C * Fast Fourier Transformation
C * ARE(N) - real part of function and its FT
C * AIM(N) - imaginary part of function and its FT
C * if P>0 - direct Fourier transformation
C * else   - inverse Fourier transformation
C * formal parameter is to be called by name
C * N - the power of 2
C * start point  - IN
C * step - K
0002          IMPLICIT REAL*8(A-H,O-Z)
0003          IMPLICIT INTEGER*4(I-N)
0004          DIMENSION ARE(N),AIM(N)
0005          N1=N/2
0006          MM=N1/2
0007          N2=N1+K
0008          J=IN
0009          JJ=J
0010        1 J=J+K
0011          IF(J-N1)2,2,10
0012        2 II=JJ+N1
0013          R=ARE(J)
0014          ARE(J)=ARE(II)
```

```
0015          ARE(II)=R
0016          R=AIM(J)
0017          AIM(J)=AIM(II)
0018          AIM(II)=R
0019          J=J+K
0020          M=MM
0021        3 IF(JJ-M)5,5,4
0022        4 JJ=JJ-M
0023          M=M/2
0024          GO TO 3
0025        5 JJ=JJ+M
0026          IF(JJ-J)1,1,6
0027        6 R=ARE(J)
0028          ARE(J)=ARE(JJ)
0029          ARE(JJ)=R
0030          R=AIM(J)
0031          AIM(J)=AIM(JJ)
0032          AIM(JJ)=R
0033          I=J+N2
0034          II=JJ+N2
0035          R=ARE(I)
0036          ARE(I)=ARE(II)
0037          ARE(II)=R
0038          R=AIM(I)
0039          AIM(I)=AIM(II)
0040          AIM(II)=R
0041          GO TO 1
0042       10 I=K
0043          T=3.14159265359
0044          IF(P)13,17,11
0045       11 T=-T
0046       13 P=-T
0047       14 SI=0.
0048          CO=1.
0049          S=DSIN(T)
0050          C=DCOS(T)
0051          T=0.5*T
0052          II=I
0053          I=I+I
0054          DO 16 M=IN,II,K
0055          DO 15 J=M,N,I
0056          JJ=J+II
0057          A=ARE(JJ)
0058          B=AIM(JJ)
```

```
0059           R=A*CO-B*SI
0060           ARE(JJ)=ARE(J)-R
0061           ARE(J)=ARE(J)+R
0062           R=B*CO+A*SI
0063           AIM(JJ)=AIM(J)-R
0064        15 AIM(J)=AIM(J)+R
0065           R=C*CO-S*SI
0066           SI=C*SI+S*CO
0067        16 CO=R
0068           IF(I-N)14,17,17
0069        17 RETURN
0070           END
```

V. Program for solving two-dimensional integral equations of convolution type

```
0001           SUBROUTINE PTITR(AK,UO,ALIM,N1,N2,Z,
              *DL,HH,C1,ALPHA,AN,U,NU,IMAX,IERR)
0002           IMPLICIT REAL*8(A-H,O-Z)
0003           IMPLICIT INTEGER*4(I-N)
0004           DIMENSION U(NU),UO(N1,N2)
0005           DIMENSION Z(N1,N2),ALIM(12)
0006           EXTERNAL AK
      C * work array mapping
      C * name    length   what
      C * ARE     N1*N2    real part of the FT of the kernel
      C * AMIN    N1*N2    imaginary part of the FT of the kernel
      C * URE     N1*N2    real part of the FT of the righthand side
      C * UIM     N1*N2    imaginary part of the FT of the rhs.
      C * ZIM     N1*N2    imaginary part of the solution
      C * W1, W2 N1, N2    stabilizer
0007           IP=1
      C * IP - start/continue mode flag
      C * start mode IP=+1
      C * reenter mode via PTIKRE IP=-1
0008           EPRO=0.
0009           IF(C1.GT.1.)EPRO=(C1-1.)*DL
      C * EPRO - accuracy of the discrepancy equation
      C * if EPRO=0 then the extremal is only computed
      C *                         with the specified ALFA
0010      100 CONTINUE
      C * calculating the support of the solution
0011           ALIM(1)=ALIM(5)-.5*(ALIM(9)+ALIM(10))
0012           ALIM(2)=ALIM(6)-.5*(ALIM(9)+ALIM(10))
```

```
0013            ALIM(3)=ALIM(7)-.5*(ALIM(11)+ALIM(12))
0014            ALIM(4)=ALIM(8)-.5*(ALIM(11)+ALIM(12))
        C * T1, T2 - periods
0015            T1=ALIM(6)-ALIM(5)
0016            T2=ALIM(8)-ALIM(7)
        C * initializing arrays
0017            NAR=1
0018            NQU=N1*N2
0019            NAI=NAR+NQU
0020            NW1=NAI+NQU
0021            NW2=NW1+N1
0022            NZI=NW2+N2
0023            NUR=NZI+NQU
0024            NUI=NUR+NQU
0025            NMAX=NUI+NQU
        C * check the length of the work array
0026            IF(NMAX-1.GT.NU) GO TO 64
0027            IF(IP.EQ.-1) GO TO 101
        C * make up righthand side
0028            CALL PTICR1(UO,UO,U(NUR),NQU,ODO)
0029            CALL PTICR2(U(NUI),ODO,NQU)
0030        101 CONTINUE
        C * solving the equation
0031            CALL PTIKR3(AK,U(NAR),U(NAI),Z,U(NZI),
               *U(NUR),U(NUI),U(NW1),U(NW2),N1,N2,
               *ALPHA,ALIM,AN,OM,T1,T2,DSQRT(DL),
               *DSQRT(HH),IP,EPRO,IQ,IMAX,IERR)
0032            GOTO 999
0033         64 CONTINUE
        C * work array is too short
0034            IERR=64
0035            GOTO 999
        C * PTITRE entry for repeatedly solving the
        C * equation with the same kernel and righthand side
0036            ENTRY PTITRE
0037            IP=-1
0038            GOTO 100
0039        999 CONTINUE
0040            RETURN
0041            END


0001            SUBROUTINE PTIKR3(AK,ARE,AIM,ZRE,ZIM,
               *URE,UIM,W1,W2,N1,N2,ALP,ALIM,BETA,RO,
```

```
            *T1,T2,DEL,HH,IPAR,EPRO,IQ,IMAX,IERR)
0002         IMPLICIT REAL*8(A-H,O-Z)
0003         IMPLICIT INTEGER*4(I-N)
0004         DIMENSION ARE(N1,N2),ZRE(N1,N2),
            *URE(N1,N2),AIM(N1,N2),ZIM(N1,N2),
            *UIM(N1,N2),W1(N1),W2(N2),ALIM(12)
0005         H1=T1/N1
0006         H2=T2/N2
0007         HA=H1/N1*H2/N2
      C * check if it is the first entry
0008         IF(IPAR.EQ.-1) GOTO 2
      C * setting kernel of the equation
0009         DO 1 I=1,N1
0010         DO 1 J=1,N2
0011         ARG1=(I-N1/2-1)*H1+0.5*(ALIM(9)+ALIM(10))
0012         ARG2=(J-N2/2-1)*H2+0.5*(ALIM(11)+ALIM(12))
0013         ARE(I,J)=AK(ARG1,ARG2)*H1*H2
0014         AIM(I,J)=0.
0015         IF(ARG1.LT.ALIM(9).OR.ARG1.GT.ALIM(10).
            OR.ARG2.LT.ALIM(11).OR.ARG2.GT.ALIM(12))
            *                            ARE(I,J)=0.
0016         W1(I)=(2.0/H1*DSIN(3.14159265359D0/N1*(I-1)))**2
0017         W2(J)=(2.0/H2*DSIN(3.14159265359D0/N2*(J-1)))**2
0018       1 CONTINUE
      C * Fourier transformation of the kernel and righthand side
0019         P=1.0
0020         CALL FTFTC(ARE,AIM,N1,N2,P)
0021         P=1.0
0022         CALL FTFTC(URE,UIM,N1,N2,P)
0023       2 CONTINUE
      C * find ALP so that PO(ALP) > 0
0024         IQ=0
0025      77 CONTINUE
0026         IQ=IQ+1
      C * calculating discrepancies
0027         F1=0.
0028         F2=0.
0029         DO 44 M=1,N1
0030         DO 44 N=1,N2
0031         U2=URE(M,N)**2+UIM(M,N)**2
0032         A2=ARE(M,N)**2+AIM(M,N)**2
0033         X=1.+(W1(M)+W2(N))**2
0034         IF(A2.EQ.0.)  GOTO 42
0035         BA=X/(A2+ALP*X)
```

```
0036           AB=1.-ALP*BA
0037           C1=HA*U2*(BA*ALP)**2
0038           C2=HA*U2*AB*BA
0039           F1=F1+C1
0040           F2=F2+C2
0041           GOTO 44
0042       42 F1=F1+HA*U2
0043       44 CONTINUE
      C * calculating generalized discrepancy
      C * to check if RO(ALP) > 0
0044           BETA=F1
0045           RO=F1-(DEL+HH*DSQRT(F2))**2
      C * if EPRO=0 then exit
0046           IF(EPRO.EQ.0.) GOTO 10
      C * if ALP=0.0 then exit
0047           IF(ALP.EQ.0.) GOTO 68
0048           IF(RO.GT.0.) GOTO 33
      C * cannot find initial regularization parameter
      C * RO(ALP) > 0
0049           IF(IQ.GT.IMAX) GOTO 65
0050           ALP=2.*ALP
0051           GOTO 77
      C * starting the Newton method
0052       33 CONTINUE
0053           IQ=0
0054        3 CONTINUE
0055           IQ=IQ+1
      C * calculating discrepancies
0056           F1=0.
0057           F2=0.
0058           F3=0.
0059           DO 4 M=1,N1
0060           DO 4 N=1,N2
0061           U2=URE(M,N)**2+UIM(M,N)**2
0062           A2=ARE(M,N)**2+AIM(M,N)**2
0063           X=1.+(W1(M)+W2(N))**2
0064           IF(A2.EQ.0.) GOTO 41
0065           BA=X/(A2+ALP*X)
0066           AB=1.-ALP*BA
0067           C1=HA*U2*(BA*ALP)**2
0068           C2=HA*U2*AB*BA
0069           C3=2.*C1*AB
0070           F1=F1+C1
0071           F2=F2+C2
```

```
0072            F3=F3+C3
0073            GOTO 4
0074      41 F1=F1+HA*U2
0075       4 CONTINUE
      C * discrepancy -BETA
      C * W21-norm of the solution -ZNOR,
      C * generalized discrepancy -RO
      C * and its derivative - DR
0076            BETA=F1
0077            ZNOR=DSQRT(F2)
0078            RO=BETA-(DEL+HH*ZNOR)**2
0079            IF(ALP.EQ.O.)  GOTO 68
0080            DR=-F3*ALP-(DEL+HH*ZNOR)*HH*F3/ZNOR
      C * accuracy achieved
0081            IF(DABS(RO).LT.EPRO) GOTO 10
      C * switch to the chord method
0082            IF(RO.LT.O.)  GOTO 61
      C * calculating new argument
      C * Newton method formulas
0083            DQ=-RO/DR
0084            A1=ALP
0085            R1=RO
0086            ALP=1./(1./ALP+DQ)
0087            IF(IQ.GE.IMAX) GOTO 66
0088            GOTO 3
      C * chord method
0089      61 CONTINUE
      C * changing interval if RO < 0
0090       6 CONTINUE
0091            AO=ALP
0092            RO=RO
0093       7 CONTINUE
0094            IQ=IQ+1
      C * getting new argument by chord method
0095            ALP=AO*A1*(RO-R1)/(AO*RO-A1*R1)
      C * calculating discrepancy
0096            F1=0.
0097            F2=0.
0098            DO 9 M=1,N1
0099            DO 9 N=1,N2
0100            U2=URE(M,N)**2+UIM(M,N)**2
0101            A2=ARE(M,N)**2+AIM(M,N)**2
0102            X=1.+(W1(M)+W2(N))**2
0103            IF(A2.EQ.O.)  GOTO 91
```

```
0104          BA=X/(A2+ALP*X)
0105          AB=1.-ALP*BA
0106          C1=HA*U2*(BA*ALP)**2
0107          C2=HA*U2*AB*BA
0108          F1=F1+C1
0109          F2=F2+C2
0110          GOTO 9
0111       91 F1=F1+HA*U2
0112        9 CONTINUE
    C * getting discrepancy, generalized discrepancy
    C * and norm of the solution
0113          BETA=F1
0114          ZNOR=DSQRT(F2)
0115          RO=BETA-(DEL+HH*ZNOR)**2
0116          IF(ALP.EQ.0.)  GOTO 68
    C * accuracy achieved
0117          IF(DABS(RO).LT.EPRO) GOTO 11
0118          IF(IQ.EQ.IMAX) GOTO 67
0119          IF(RO.LT.0.)  GOTO 6
    C * changing interval if RO > 0
0120          A1=ALP
0121          R1=RO
0122          GOTO 7
0123       65 CONTINUE
    C * cannot find ALP so that RO(ALP)>0
0124          IERR=65
0125          GOTO 999
0126       66 CONTINUE
    C * IMAX Newton iterations made
0127          IERR=66
0128          GOTO 999
0129       67 CONTINUE
    C * IMAX chord iterations made
0130          IERR=67
0131          GOTO 999
0132       68 CONTINUE
    C * ALP=0.0 is specified or found
0133          IERR=68
0134          GOTO 999
0135       11 CONTINUE
    C * solution is found by chord method
0136          IERR=1
0137          GOTO 999
0138       10 CONTINUE
```

```
         C * solution is found by Newton's method
0139         IERR=0
0140     999 CONTINUE
         C * getting Fourier transform of the solution
0141         SSI=1.0
0142         DO 12 M=1,N1
0143         SSI=-SSI
0144         DO 12 N=1,N2
0145         SSI=-SSI
0146         ZZ=N1*N2*(ARE(M,N)**2+AIM(M,N)**2+
             +ALP*(1.0+(W1(M)+W2(N))**2))
0147         IF(ZZ.NE.0.)  GOTO 111
0148         ZRE(M,N)=0.
0149         ZIM(M,N)=0.
0150         GOTO 12
0151     111 ZRE(M,N)=SSI*(ARE(M,N)*URE(M,N)+
             +AIM(M,N)*UIM(M,N))/ZZ
0152         ZIM(M,N)=SSI*(ARE(M,N)*UIM(M,N)-
             -AIM(M,N)*URE(M,N))/ZZ
0153      12 CONTINUE
         C * inverse Fourier transformation
0154         P=-1.0
0155         CALL FTFTC(ZRE,ZIM,N1,N2,P)
0156         RETURN
0157         END


0001         SUBROUTINE FTFTC(ARE,AIM,N1,N2,P)
         C * Fast two-dimensional Fourier transformation
         C * if P > 0 - direct transformation
         C * if P < 0 - inverse transformation
         C * P - must be called by name
0002         IMPLICIT REAL*8(A-H,O-Z)
0003         IMPLICIT INTEGER*4(I-N)
0004         DIMENSION ARE(N1,N2),AIM(N1,N2)
0005         DO 1 I=1,N1
0006         CALL FTF1C(ARE,AIM,N1*N2,I,N1,P)
0007       1 CONTINUE
0008         DO 2 J=1,N2
0009         CALL FTF1C(ARE(1,J),AIM(1,J),N1,1,1,P)
0010       2 CONTINUE
0011         RETURN
0012         END
```

```
0001          SUBROUTINE FTF1C(XARE,XAIM,N,IN,K,P)
0002          IMPLICIT REAL*8(A-H,O-Z)
0003          IMPLICIT INTEGER*4(I-N)
0004          DIMENSION XARE(N),XAIM(N)
0005          N1=N/2
0006          MM=N1/2
0007          N2=N1+K
0008          J=IN
0009          JJ=J
0010        1 J=J+K
0011          IF(J-N1)2,2,10
0012        2 II=JJ+N1
0013          R=XARE(J)
0014          XARE(J)=XARE(II)
0015          XARE(II)=R
0016          R=XAIM(J)
0017          XAIM(J)=XAIM(II)
0018          XAIM(II)=R
0019          J=J+K
0020          M=MM
0021        3 IF(JJ-M)5,5,4
0022        4 JJ=JJ-M
0023          M=M/2
0024          GOTO 3
0025        5 JJ=JJ+M
0026          IF(JJ-J)1,1,6
0027        6 R=XARE(J)
0028          XARE(J)=XARE(JJ)
0029          XARE(JJ)=R
0030          R=XAIM(J)
0031          XAIM(J)=XAIM(JJ)
0032          XAIM(JJ)=R
0033          I=J+N2
0034          II=JJ+N2
0035          R=XARE(I)
0036          XARE(I)=XARE(II)
0037          XARE(II)=R
0038          R=XAIM(I)
0039          XAIM(I)=XAIM(II)
0040          XAIM(II)=R
0041          GOTO 1
0042       10 I=K
0043          T=3.14159265359
0044          IF(P)13,17,11
```

```
0045        11 T=-T
0046        13 P=-T
0047        14 SI=0.
0048           CO=1.
0049           S=DSIN(T)
0050           C=DCOS(T)
0051           T=0.5*T
0052           II=I
0053           I=I+I
0054           DO 16 M=IN,II,K
0055           DO 15 J=M,N,I
0056           JJ=J+II
0057           A=XARE(JJ)
0058           B=XAIM(JJ)
0059           R=A*CO-B*SI
0060           XARE(JJ)=XARE(J)-R
0061           XARE(J)=XARE(J)+R
0062           R=B*CO+A*SI
0063           XAIM(JJ)=XAIM(J)-R
0064        15 XAIM(J)=XAIM(J)+R
0065           R=C*CO-S*SI
0066           SI=C*SI+S*CO
0067        16 CO=R
0068           IF(I-N)14,17,17
0069        17 RETURN
0070           END
```

VI. Program for solving Fredholm integral equations of the first kind on the sets of monotone and (or) convex functions. The method of the conditional gradient

```
0001            SUBROUTINE PTIGR(AK,U0,X1,X2,Y1,Y2,N,M,
               *Z,AN2,ITER,DL,IMAX,C1,C2,IC,R,NR,IERR)
          C * Solving integral equations of the first kind
          C * using the conditional gradient method
          C *  DISPATCHER
          C * AK - kernel function
          C * Work array mapping
          C * name       length      what it is
          C *   A:        N*M      operator matrix
          C *   H:        N        descent direction
          C *   G:        N        gradient
          C *   U:        M        operator value
          C *   U1:       M           work array
```

```
       C *
       C *  NR=N*M+2(N+M)
0002         IMPLICIT REAL*8(A-H,O-Z)
0003         IMPLICIT INTEGER*4(I-N)
0004         EXTERNAL AK
0005         EXTERNAL PTICRO,PTIGR1
0006         DIMENSION UO(M),Z(N),R(NR)
       C * get arrays indexes
0007         NA=1
0008         NH=N*M+1
0009         NG=NH+N
0010         NU=NG+N
0011         NU1=NU+M
0012         NMAX=NU1+M
0013         IF(NMAX-1.GT.NR)GOTO 64
       C * getting operator matrix
0014         CALL PTICRO(AK,R(NA),X1,X2,Y1,Y2,N,M)
       C * minimizing discrepancy
0015         CALL PTIGR1(R(NA),Z,UO,C1,C2,IC,N,M,
            *ITER,DL*(M-1.)/(Y2-Y1),ODO,IMAX,AN2,
            *Z,R(NU),R(NU1),R(NH),R(NG),IERR)
0016         AN2=AN2*(Y2-Y1)/(M-1.)
0017         GOTO 999
0018    64   CONTINUE
       C * work array is too short
0019         IERR=64
0020         GOTO 999
0021    999  CONTINUE
0022         RETURN
0023         END


0001         SUBROUTINE PTIGR1(A,ZO,UO,C1,C2,IC,N,M,
            *ITER,DL2,ANGRD,IMAX,AN2,
            *Z,U,U1,H,G,IERR)
       C * conditional gradient method for
       C * minimizing the discrepancy
       C * A(M,N) - operator matrix
       C * ZO(N) - initial approximation
       C * UO(M) - righthand side of the equation
       C * C1,C2 - constraints restricting the solution
       C * IC - type of set of correctness
       C * ITER - number of iterations made
       C * DL2 - discrepancy level to stop iterations
```

```
       C * ANGRD - iteration stop level for norm of the gradient
       C * IMAX - maximum number of iterations
       C * Z(N) - solution found
       C * U(M) - operator value on the solution
       C * AN2 - discrepancy value for the solution
       C * G(N),H(N) - work arrays
       C * U1(M) - work array for one-dimensional minimization
0002        IMPLICIT REAL*8(A-H,O-Z)
0003        IMPLICIT INTEGER*4(I-N)
0004        EXTERNAL PTIGR2,PTICR1,PTICR3
0005        EXTERNAL PTICR5,PTICR4,PTICR6
0006        DIMENSION U1(M),U(M),G(N),H(N),Z(N)
0007        DIMENSION A(M,N),Z0(N),U0(M)
0008        ITER=0
0009        ALMAX=1.
0010        CALL PTICR1(Z0,Z0,Z,N,ODO)
0011        CALL PTICR3(A,Z,U,N,M)
0012        CALL PTICR5(U,U0,M,AN20)
0013        CALL PTICR4(G,U,U0,A,N,M)
0014        CALL PTICR6(G,G,N,ANGR)
0015        IF(AN20.LE.DL2.OR.ANGR.LE.ANGRD.OR.
            ITER.GE.IMAX)GOTO 20
       C * start of iterations
0016   14   CONTINUE
0017        ITER=ITER+1
0018        CALL PTIGR2(H,G,N,C1,C2,IC)
0019        CALL PTICR1(Z,H,H,N,-1D0)
0020        CALL PTICR0(A,Z,G,U1,H,ALMAX,AL,N,M,0.D0,0.D0,IED)
0021        CALL PTICR1(Z,H,Z,N,-AL)
0022        CALL PTICR3(A,Z,U,N,M)
0023        CALL PTICR4(G,U,U0,A,N,M)
0024        CALL PTICR6(G,G,N,ANGR)
0025        CALL PTICR5(U,U0,M,AN2)
0026        IF(AN2.LE.DL2.OR.ANGR.LE.ANGRD.OR.ITER.GE.IMAX)
            +      GOTO 20
0027        IF(AN2.GE.AN20)GOTO 21
       C * storing discrepancy
0028        AN20=AN2
0029        GOTO 14
0030   20   CONTINUE
       C * normal end
0031        IERR=0
0032        GOTO 999
0033   21   CONTINUE
```

```
         C * discrepancy hasn't decreased
0034            IERR=1
0035      999 CONTINUE
0036            RETURN
0037            END


0001            SUBROUTINE PTIGR2(TOP,G,N,C1,C2,IC)
         C * finding optimal vertex
         C * G - gradient array
         C * TOP - vertex found, array
         C * C1,C2 - restriction constants
         C * IC=1 - monotone functions
         C * IC=2 - monotone convex functions
         C * IC=3 - convex functions
         C * IC=-1,C1=0. - functions of finite variation
         C *       Var F <= 2*C2
0002            IMPLICIT REAL*8(A-H,O-Z)
0003            IMPLICIT INTEGER*4(I-N)
0004            EXTERNAL PTICR1,PTICR2
0005            DIMENSION TOP(N),G(N)
0006            AL=0.
0007            S1=0.
0008            S2=0.
0009            B=1./(N-1.)
0010            A=1.-B
0011            SS=+1.
         C * in K1 - number of the current optimal vertex
0012            K1=0
0013            ICM=IABS(IC)
0014            IF(ICM.EQ.1)GOTO 101
0015            DO 1 I=2,N
0016            S2=S2+G(I)*A
0017      1     A=A-B
         C * in S2 - value of the linear functional
         C * without the left point
0018      101 CONTINUE
0019            DO 2 K=1,N
0020            SG=S1+S2+G(K)
         C * SG - value of the functional
0021            IF(AL.LE.SG)GOTO 3
0022            AL=SG
0023            K1=K
0024            SS=+1.
```

```
0025     3    CONTINUE
0026          IF(IC.GE.0)GOTO 8
         C * checking symmetrical vertex, if needed (IC<0)
0027          IF(AL.LE.-SG)GOTO 4
0028          AL=-SG
0029          K1=K
0030          SS=-1.
         C * SS=-1 - setting the flag:  symmetric value used
0031     4    CONTINUE
0032     8    CONTINUE
         C * in S1 - first half of the functional
         C * and in S2 - the second
0033          IF(ICM.EQ.1.OR.ICM.EQ.2)S1=S1+G(K)
0034          IF(ICM.EQ.3)S1=(S1+G(K))*(K-1.)/K
0035          IF(K.GE.N-1.OR.ICM.EQ.1)GOTO 7
0036          S2=S2*(N-K)/(N-K-1.)-G(K+1)
0037          GOTO 6
0038     7    S2=0.
0039     6    CONTINUE
0040     2    CONTINUE
         C * number of optimal vertex is found - K1
         C * form this vertex vector
0041          IF(K1.NE.0)GOTO 9
0042          CALL PTICR2(TOP,C1,N)
0043          GOTO 999
0044     9    CONTINUE
0045          DO 5 I=1,N
0046          GOTO (801,802,803),ICM
0047     801  IF(I.LE.K1)TOP(I)=C2
0048          IF(I.GT.K1)TOP(I)=C1
0049          GOTO 5
0050     802  IF(I.LE.K1)TOP(I)=C2
0051          IF(I.GT.K1)
         *           TOP(I)=TOP(I-1)-(C2-C1)/(N-K1)
0052          GOTO 5
0053     803  IF(I.LE.K1.AND.K1.NE.1)
         *     TOP(I)=C1+(C2-C1)/(K1-1.)*(I-1.)*2.
0054          IF(I.GE.K1.AND.K1.NE.N)
         *     TOP(I)=C1+(C2-C1)/(N-K1)*(N-I+0.)*2.
0055     5    CONTINUE
0056          IF(IC.GE.0)GOTO 999
0057          CALL PTICR1(TOP,TOP,TOP,N,SS-1.)
0058          GOTO 999
0059     999  CONTINUE
```

```
0060           RETURN
0061           END
```

VII. Program for solving Fredholm integral equations of the first kind on the sets of monotone and (or) convex functions. The method of projection of conjugate gradients

```
0001           SUBROUTINE PTILR(AK,U0,X1,X2,Y1,Y2,N,M,
              *Z,AN2,DL,ITER,IMAX,C2,IC,R,NR,IERR)
0002           IMPLICIT REAL*8(A-H,O-Z)
0003           IMPLICIT INTEGER*4(I-N)
0004           DIMENSION U0(M),Z(N),R(NR)
0005           EXTERNAL AK
0006           EXTERNAL PTICR0,PTICR1,
              *PTILRA,PTILRB,PTILR1,PTILR2
       C * work array mapping
       C * name length     what
       C *   IM:      1     nr. of active constraints for PTILR1(PTILR2)
       C *    A:    N*M     operator matrix
       C *    C:    N*N     (C*X,X)
       C *    D:      N     (D,X)
       C *  CON:   NN*N     matrix of constraints
       C *    B:     NN     vector of constraints
       C * MASK:     NN     mask of active constraints
       C *   AJ:   NN*N     matrix of active constraints
       C *    P:   NN*N     work array
       C *   PI:    N*N     projector (work)
       C *   GR:      N     work array
       C *    W: MAX(NN,N,M)   work arrays
       C *   P1:      N         work array
       C *
       C * NR>=3*N*M+3*N*N+8*N+MAX(M,N+1)+3
       C *
       C * ICONT=0 - first call
       C * ICONT=1 - next call for continuing
0007           ICONT=0
0008     101   CONTINUE
       C * number of constraints NN
0009           NN=N
0010           IF(IC.EQ.1)NN=N+1
       C * initializing arrays
0011           NA=2
0012           NC=NA+N*M
0013           ND=NC+N*N
```

```
0014          NCON=ND+N
0015          NB=NCON+N*NN
0016          NMASK=NB+NN
0017          NAJ=NMASK+NN
0018          NP=NAJ+NN*N
0019          NPI=NP+NN*N
0020          NGR=NPI+N*N
0021          NW=NGR+N
0022          NP1=NW+MAXO(NN,M,N)
0023          NMAX=NP1+N
0024          IF(NMAX-1.GT.NR)GOTO 64
0025          IF(ICONT.EQ.1)GOTO 100
       C * if ICONT doesn't equal 1 get operator matrix
0026          CALL PTICRO(AK,R(NA),X1,X2,Y1,Y2,N,M)
       C * transforming discrepancy functional to the form
       C * (C*X,X)+(D,X)+E
0027          CALL PTILRA(R(NA),UO,R(NC),R(ND),N,M)
       C * getting matrix of constraints
0028          CALL PTILRB(R(NCON),R(NB),N,NN,IC,C2,IE)
0029          IF(IE.NE.0)GOTO 65
       C * minimizing discrepancy
0030          CALL PTILR1(M,MR,R(NMASK),R(NA),R(NAJ),
              *R(NCON),R(NP),UO,R(NPI),R(NB),R(NGR),
              *R(NW),R(NP1),.TRUE.,IMAX,ITER,R(NC),Z,
              *R(ND),N,NN,AN2,DL*(M-1.)/(Y2-Y1),
              *0.0D0,IERR)
0031   102    CONTINUE
0032          AN2=AN2*(Y2-Y1)/(M-1.)
0033          GOTO 999
0034          ENTRY PTILRE
       C * entry to continue
0035          ICONT=1
0036          GOTO 101
0037   100    CONTINUE
       C * call continue program
0038          CALL PTILR2(M,MR,R(NMASK),R(NA),R(NAJ),
              *R(NCON),R(NP),UO,R(NPI),R(NB),R(NGR),
              *R(NW),R(NP1),.TRUE.,IMAX,ITER,R(NC),Z,
              *R(ND),N,NN,AN2,DL*(M-1.)/(Y2-Y1),
              *0.0D0,IERR)
0039          GOTO 102
0040   64     CONTINUE
       C * work array is too short
0041          IERR=64
```

```
0042          GOTO 999
0043    65    CONTINUE
      C * invalid correctness set
0044          IERR=65
0045          GOTO 999
0046   999    CONTINUE
0047          RETURN
0048          END


0001          SUBROUTINE PTILR5(A,AJ,B1,RMASK,N,M,NN,
             *GR,PI,P,P1,C,X,D,K,IED,DGR)
      C * minimization of quadratic functional
      C * on the set defined by the active constraints AJ
      C * numbers of active constraints in RMASK,
      C * start point - X, end point - X,
      C * projector - PI.
      C * IED - return code:
      C *      0 - new face achieved,
      C *      1 - minimum found
      C *      2 - gradient norm is too small
0002          IMPLICIT REAL*8(A-H,O-Z)
0003          IMPLICIT INTEGER*4(I-N)
0004          DIMENSION AJ(NN,N),B1(NN),RMASK(NN),
             *GR(N),PI(N,N),P(N),P1(N),C(N,N),
             *X(N),A(NN,N),D(N)
0005          EXTERNAL PTICR1,PTICR2,
             *PTICR3,PTICR6,PTILR7
      C * K - iteration count on the subspace
0006          K=0
0007          AN1=1.
      C * when K=0 the descent direction is 0.
0008          CALL PTICR2(P,0.0D0,N)
      C * loop of the conjugate gradient method
      C * check minimum by number of iterations
0009    11 IF(K.EQ.N-M) GOTO 14
      C * get gradient, if its norm < DGR then exit
0010          CALL PTILR7(X,C,N,GR,D)
0011          CALL PTICR6(GR,GR,N,R)
0012          IF(R.LT.DGR) GOTO 17
      C * P1 - antigradient projection
0013          CALL PTICR3(PI,GR,P1,N,N)
0014          CALL PTICR1(P1,P1,P1,N,-2.0D0)
      C * AN - projection norm
```

```
        C * AN2- conjugation coefficient
0015            CALL PTICR6(P1,P1,N,AN)
0016            AN2=AN/AN1
        C * P - new conjugate (descent) direction
0017            CALL PTICR1(P1,P,P,N,AN2)
        C * R:=(P,GR)
0018            CALL PTICR6(P,GR,N,R)
        C * R1:=(CP,P)
0019            R1=0.0
0020            DO 4 I=1,N
0021            R2=0.0
0022            DO 5 J=1,N
0023            R2=R2+C(I,J)*P(J)
0024          5 CONTINUE
0025            R1=R1+P(I)*R2
0026          4 CONTINUE
0027            IF(R1.EQ.0.0) GOTO 14
        C * optimal step =-0.5R/R1
0028            ALPHA=-R/R1/2.
        C * finding the nearest plane
0029            ALM=ALPHA+1.0
0030            NP=0
0031        701 DO 6 I=1,NN
        C * don't testing active constraints
0032            IF(RMASK(I).EQ.1) GOTO 6
0033            R=0.0
0034            R1=0.0
0035            DO 7 J=1,N
0036            R=R+A(I,J)*X(J)
0037            R1=R1+A(I,J)*P(J)
0038          7 CONTINUE
        C * R1 > 0
0039            IF(R1.LE.0.0) GOTO 6
        C * ALP - distance to plane
0040            ALP=(B1(I)-R)/R1
        C * finding minimal one
0041            IF(ALM.LE.ALP) GOTO 6
0042            ALM=ALP
        C * NP - constraint number
0043            NP=I
0044          6 CONTINUE
        C * if ALPHA > ALM - new constraint
0045            IF(ALPHA.GE.ALM) GOTO 12
        C * if ALPHA < 0 - minimum is found,
```

```
       C * else - to the next step
0046          IF(ALPHA.LE.0.0) GOTO 14
0047          CALL PTICR1(X,P,X,N,ALPHA)
0048          K=K+1
0049          AN1=AN
0050          GOTO 11
0051       12 ALPHA=ALM
       C * going to the new point on the boundary
0052          CALL PTICR1(X,P,X,N,ALPHA)
       C * including new constraint into the restriction mask
0053          RMASK(NP)=1
0054          M=M+1
0055          IED=0
0056          GOTO 16
0057       14 IED=1
0058          GOTO 16
0059       17 CONTINUE
0060          IED=2
0061       16 CONTINUE
0062          RETURN
0063          END


0001          SUBROUTINE PTILR7(X,C,N,GR,D)
       C * calculating the gradient of
       C * the quadratic functional
0002          IMPLICIT REAL*8(A-H,O-Z)
0003          IMPLICIT INTEGER*4(I-N)
0004          DIMENSION C(N,N),X(N),GR(N),D(N)
       C * GR:=2LX+D
0005          DO 1 I=1,N
0006          R=0.0
0007          DO 2 J=1,N
0008          R=R+(C(I,J)+C(J,I))*X(J)
0009       2 CONTINUE
0010          GR(I)=R+D(I)
0011       1 CONTINUE
0012          RETURN
0013          END


0001          SUBROUTINE PTILR4(AJ,P,PI,M,N,NN)
       C * getting projector
0002          IMPLICIT REAL*8(A-H,O-Z)
```

```
0003            IMPLICIT INTEGER*4(I-N)
0004            DIMENSION AJ(NN,N),P(NN,N),PI(1)
0005            EXTERNAL PTILR0,PTICR2
0006            IF(M.EQ.0) GOTO 1
      C * PI:=AJ*AJ'
0007            DO 2 I=1,M
0008            DO 2 J=1,I
0009            R=0.0
0010            DO 3 K=1,N
0011            R=R+AJ(I,K)*AJ(J,K)
0012          3 CONTINUE
0013            L=M*(I-1)+J
0014            PI(L)=R
0015            L=M*(J-1)+I
0016            PI(L)=R
0017          2 CONTINUE
      C * PI:=INV(PI)
0018            CALL PTILR0(PI,M)
      C * P:=-PI*AJ
0019            DO 4 I=1,M
0020            DO 4 J=1,N
0021            R=0.0
0022            DO 5 K=1,M
0023            L=M*(I-1)+K
0024            R=R+PI(L)*AJ(K,J)
0025          5 CONTINUE
0026            P(I,J)=-R
0027          4 CONTINUE
      C * PI:=AJ'*P
0028            DO 6 I=1,N
0029            DO 6 J=1,I
0030            R=0.0
0031            DO 7 K=1,M
0032            R=R+AJ(K,I)*P(K,J)
0033          7 CONTINUE
0034            L=N*(I-1)+J
0035            PI(L)=R
0036            L=N*(J-1)+I
0037            PI(L)=R
0038          6 CONTINUE
0039            GOTO 10
0040          1 CONTINUE
      C * if there are no active constraints then PI:=0, P:=0
0041            CALL PTICR2(PI,0.0D0,N*N)
```

```
0042          CALL PTICR2(P,0.0D0,N*NN)
        C * PI:=E+PI
0043       10 DO 9 I=1,N
0044          L=N*(I-1)+I
0045          PI(L)=PI(L)+1.0
0046        9 CONTINUE
0047          RETURN
0048          END


0001          SUBROUTINE PTILR3(AJ,A,N,NN,RMASK)
        C * geting active constraints by the mask
0002          IMPLICIT REAL*8(A-H,O-Z)
0003          IMPLICIT INTEGER*4(I-N)
0004          DIMENSION AJ(NN,N),A(NN,N),RMASK(NN)
0005          K=0
0006          DO 1 I=1,NN
0007          IF(RMASK(I).EQ.0) GOTO 1
0008          K=K+1
0009          DO 2 J=1,N
0010          AJ(K,J)=A(I,J)
0011        2 CONTINUE
0012        1 CONTINUE
0013          RETURN
0014          END


0001          SUBROUTINE PTILR6(A,X,P,RMASK,
             *GR,N,M,NN,C,D,IED)
        C * excluding active constraints
        C * IED - return code:
        C *      0 - nothing to exclude
        C *      1 - something is excluded
0002          IMPLICIT REAL*8(A-H,O-Z)
0003          IMPLICIT INTEGER*4(I-N)
0004          DIMENSION P(NN,N),GR(N),RMASK(NN)
0005          DIMENSION A(NN,N),X(N),C(N,N),D(N)
0006          EXTERNAL PTILR7
0007          CALL PTILR7(X,C,N,GR,D)
0008          K=0
0009          IED=0
0010          AL=1.0
0011          DO 1 I=1,NN
0012          IF(RMASK(I).EQ.0) GOTO 1
```

```
         C * getting shadow parameter for active constraints
0013          K=K+1
0014          R=0.0
0015          DO 2 J=1,N
0016          R=R+P(K,J)*GR(J)
0017        2 CONTINUE
         C * getting minimal shadow parameter
0018          IF (R.GE.AL) GOTO 1
0019          AL=R
0020          NNN=I
0021        1 CONTINUE
         C * if minimal shadow parameter > 0 then exit
         C * else excluding NNN-th constraint
0022          IF (AL.GE.0.00000) GOTO 3
0023          RMASK(NNN)=0
0024          IED=1
0025          M=M-1
0026        3 RETURN
0027          END


0001          SUBROUTINE PTILR1(MN,M,RMASK,AQ,AJ,A,P,
              *U,PI,B,GR,W,P1,WORK,ICM,ICI,C,X,D,N,NN,
              *R,DEL,DGR,IEND)
         C *      dispatcher
         C * WORK=.TRUE.- working with the matrix A being set
         C * ICI - large loops counter
         C * ICM - maximum number of large loops
0002          IMPLICIT REAL*8(A-H,O-Z)
0003          IMPLICIT INTEGER*4(I-N)
0004          DIMENSION A(NN,N),RMASK(NN),AJ(NN,N),
              *P(NN,N),PI(N,N),B(NN),GR(N),W(MN),
              *P1(N),C(N,N),X(N),D(N),AQ(MN,N),U(MN)
0005          LOGICAL WORK
0006          EXTERNAL PTICI2,PTICR3,PTICR5,
              *PTILR3,PTILR4,PTILR5,PTILR6
0007          CALL PTICR2(RMASK,0.0D0,NN)
         C * M - number of active constraints
0008          M=0
0009          ENTRY PTILR2
         C * starting large loops iterations
0010          ICI=0
0011        2 CONTINUE
0012          IF(ICI.EQ.ICM) GOTO 12
```

```
0013          ICI=ICI+1
0014          IF(.NOT.WORK) GOTO 101
      C * calculating discrepancy - R1,
      C * if it is less < DEL then exit
0015          CALL PTICR3(AQ,X,W,N,MN)
0016          CALL PTICR5(W,U,MN,R1)
0017   101    CONTINUE
0018          IF(R1.LE.DEL) GOTO 9
      C * getting constraints matrix by the mask
0019          CALL PTILR3(AJ,A,N,NN,RMASK)
      C * getting projector
0020          CALL PTILR4(AJ,P,PI,M,N,NN)
      C * minimizing on the face
0021          CALL PTILR5(A,AJ,B,RMASK,N,M,NN,GR,
             *PI,W,P1,C,X,D,K,IED,DGR)
0022          IF(IED.EQ.0) GOTO 2
      C * if gradient norm < DGR then exit
0023          IF(IED.EQ.2) GOTO 11
      C * trying to exclude active constraint
0024          CALL PTILR6(A,X,P,RMASK,GR,N,M,NN,C,D,IED)
      C * if something was excluded then continue calculations
0025          IF(IED.EQ.1) GOTO 2
      C * normal return
      C   exact mimimum is found
0026          IEND=0
0027          GOTO 10
0028    9     CONTINUE
      C * normal return
      C * discrepancy level achieved
0029          IEND=1
0030          GOTO 10
0031    12    CONTINUE
      C * maximum large loop number achieved
0032          IEND=3
0033          GOTO 10
0034     11   CONTINUE
      C * gradient norm is too small
0035          IEND=2
0036     10   CONTINUE
0037          IF(.NOT.WORK)GOTO 102
      C * calculating discrepancy
0038          CALL PTICR3(AQ,X,W,N,MN)
0039          CALL PTICR5(W,U,MN,R)
0040    102   CONTINUE
```

```
0041          RETURN
0042          END


0001          SUBROUTINE PTILRA(AQ,U,C,D,N,MN)
       C * transforming the discrepancy functional
       C * to the form (C*X,X)+(D,X)+E
0002          IMPLICIT REAL*8(A-H,O-Z)
0003          IMPLICIT INTEGER*4(I-N)
0004          DIMENSION AQ(MN,N),U(MN),C(N,N),D(N)
       C * D:=-2A'*U
0005          DO 6 I=1,N
0006          R=0.0
0007          DO 5 J=1,MN
0008          R=R+AQ(J,I)*U(J)
0009        5 CONTINUE
0010          D(I)=-2.0*R
0011    6     CONTINUE
       C * L:=A'*A
0012          DO 7 I=1,N
0013          DO 7 J=1,I
0014          R=0.0
0015          DO 8 K=1,MN
0016          R=R+AQ(K,I)*AQ(K,J)
0017        8 CONTINUE
0018          C(I,J)=R
0019          C(J,I)=R
0020    7     CONTINUE
0021          RETURN
0022          END


0001          SUBROUTINE PTILRB(A,B,N,NN,ITASK,C,IERR)
       C * forming matrix of constraints
0002          IMPLICIT REAL*8(A-H,O-Z)
0003          IMPLICIT INTEGER*4(I-N)
0004          DIMENSION A(NN,N),B(NN)
0005          EXTERNAL PTICR2
       C * invalid constraints type set
0006          IERR=99
0007          IF(ITASK.GT.3.OR.ITASK.LT.1) GOTO 777
       C * for monotone functions we
       C * test if the constraint constant is non-zero
0008          IERR=98
```

```
0009            IF(ITASK.EQ.1.AND.C.EQ.0.0) GOTO 777
0010            IERR=0
       C * send zero to the matrix
0011            CALL PTICR2(A,0.0D0,N*NN)
0012            L=N-1
0013            GOTO (1,2,3),ITASK
0014          1 CONTINUE
       C * monotone functions
0015            DO 11 I=1,N
0016            A(I,I)=1.0
0017            A(I+1,I)=-1.0
0018            B(I+1)=0.0
0019         11 CONTINUE
0020            B(1)=C
0021            GOTO 777
0022          2 CONTINUE
       C * convex functions
0023            DO 12 I=2,L
0024            A(I,I)=-2.0
0025            A(I,I-1)=1.0
0026            A(I,I+1)=1.0
0027            B(I)=0.0
0028         12 CONTINUE
0029            A(1,1)=-1.0
0030            A(N,N)=-1.0
0031            B(1)=0.0
0032            B(N)=0.0
0033            GOTO 777
0034          3 CONTINUE
       C * monotone convex functions
0035            DO 13 I=2,L
0036            A(I,I)=-2.0
0037            A(I,I-1)=1.0
0038            A(I,I+1)=1.0
0039            B(I)=0.0
0040         13 CONTINUE
0041            A(1,1)=-1.0
0042            A(1,2)=1.0
0043            A(N,N)=-1.0
0044            B(1)=0.0
0045            B(N)=0.0
0046        777 CONTINUE
0047            RETURN
0048            END
```

```
0001            SUBROUTINE PTILRO(A,N)
       C * getting inverse matrix
       C * of a symmetric positive definite matrix A(N,N)
       C * N+1-th column of the matrix A ( A(N,N+1) )
       C * is used as work array
       C * only subdiagonal elements may be set
       C *        A(I,J), where I>=J
0002            IMPLICIT REAL*8(A-H,O-Z)
0003            IMPLICIT INTEGER*4(I-N)
0004            DIMENSION A(N,1)
0005            DO 1 I=1,N
0006            I1=I+1
0007            DO 1 J=I,N
0008            J1=J+1
0009            X=A(J,I)
0010            I2=I-1
0011            IF(I2.LT.1)GOTO 2
0012            DO 3 K1=1,I2
0013            K=I-K1
0014            X=X-A(K,J1)*A(K,I1)
0015    3       CONTINUE
0016    2       CONTINUE
0017            IF(J.NE.I) GOTO 4
0018            Y=1./DSQRT(X)
0019            A(I,I1)=Y
0020            GOTO 1
0021    4       A(I,J1)=X*Y
0022    1       CONTINUE
0023            DO 5 I=1,N
0024            I3=I+1
0025            IF(I3.GT.N)GOTO 5
0026            DO 6 J=I3,N
0027            Z=0.
0028            J1=J+1
0029            J2=J-1
0030            DO 7 K1=I,J2
0031            K=J-1-K1+I
0032    7       Z=Z-A(K,J1)*A(I,K+1)
0033            A(I,J1)=Z*A(J,J1)
0034    6       CONTINUE
0035    5       CONTINUE
0036            DO 8 I=1,N
0037            DO 8 J=I,N
0038            Z=0.
```

```
0039           J1=N+1
0040           J4=J+1
0041           DO 9 K=J4,J1
0042     9     Z=Z+A(J,K)*A(I,K)
0043           A(I,J+1)=Z
0044     8     CONTINUE
0045           DO 11 I=1,N
0046           DO 11 J=I,N
0047           A(J,I)=A(I,J+1)
0048           A(I,J)=A(I,J+1)
0049    11     CONTINUE
0050           RETURN
0051           END
```

VIII. Program for solving Fredholm integral equations of the first kind on the sets of monotone and (or) convex functions. The method of projection of conjugate gradients onto the set of vectors with nonnegative coordinates

```
0001           SUBROUTINE PTISR(AK,U0,X1,X2,Y1,Y2,N,M,
              *Z,AN2,ITER,DL,IMAX,IC,
              *R,NR,IERR)
         C * Program to solve integral equations
         C * of the first kind by the method of
         C * projection of conjugate gradients
         C * dispatcher
         C * work array mapping
         C *   name      length       what
         C *    A:        N*M       operator matrix
         C *    H:         N        descent direction
         C *    G:         N        gradient
         C *    U:         M        operator value
         C *    U1:        M        work array
         C *    S:         N        work array
         C *
         C * NR=N*M+3N+2M
0002           IMPLICIT REAL*8(A-H,O-Z)
0003           IMPLICIT INTEGER*4(I-N)
0004           DIMENSION U0(M),Z(N),R(NR)
0005           EXTERNAL AK
0006           EXTERNAL PTICR0,PTICR1,
              *PTISR1,PTISR2,PTISR3
0007           ICONT=0
         C * ICONT - start/continue mode flag
```

```
        C *    ICONT=0 - start mode
        C *    ICONT=1 - continue mode
0008    100  CONTINUE
        C * initializing arrays
0009         NA=1
0010         NH=N*M+1
0011         NG=NH+N
0012         NU=NG+N
0013         NU1=NU+M
0014         NS=NU1+M
0015         NMAX=NS+N
0016         IF(NMAX-1.GT.NR)GOTO 64
0017         IF(ICONT.EQ.1)GOTO 101
        C * getting operator matrix
0018         CALL PTICR0(AK,R(NA),X1,X2,Y1,Y2,N,M)
        C * tranformation to PI-plus
0019         CALL PTISR2(R(NA),Z,N,M,IC,R(NS))
0020         CALL PTICR1(Z,Z,R(NH),N,ODO)
0021    101  CONTINUE
        C * minimizing discrepancy
0022         CALL PTISR1(R(NA),R(NH),UO,N,M,ITER,
             *DL*(M-1.)/(Y2-Y1),ODO,IMAX,AN2,ODO,ODO,
             *Z,R(NU),R(NU1),R(NH),R(NG),R(NS),IERR)
0023         AN2=AN2*(Y2-Y1)/(M-1.)
        C * return from PI-plus to original coordinates
0024         CALL PTICR1(Z,Z,R(NH),N,ODO)
0025         CALL PTISR3(Z,N,IC,R(NS))
0026         GOTO 999
0027         ENTRY PTISRE
        C * entry to continue calculations
0028         ICONT=1
0029    64   CONTINUE
        C * work array is too short
0030         IERR=64
0031         GOTO 999
0032    999  CONTINUE
0033         RETURN
0034         END


0001         SUBROUTINE PTISR1(A,Z0,UO,N,M,ITER,DL2,
             *ANGRD,IMAX,AN2,ALF,RO,
             *Z,U,U1,H,G,PLUS,IERR)
        C * minimization of Thikhonov's functional
```

```
       C * by the conjugate gradient method
       C * in the first quadrant
0002          IMPLICIT REAL*8(A-H,O-Z)
0003          IMPLICIT INTEGER*4(I-N)
0004          DIMENSION UO(M),ZO(N),Z(N),U(M),
              *A(M,N),H(N),U1(M),G(N),PLUS(N)
0005          EXTERNAL PTICI2,PTICR3,PTICR5,PTICR4
0006          EXTERNAL PTICR6,PTICR7,PTICR1
0007          EXTERNAL PTICR9,PTICR8
0008          ITERBN=0
0009          IEND=0
0010          ITER=0
0011          JCH=-1
0012          CALL PTICR1(ZO,ZO,Z,N,ODO)
0013          CALL PTICR3(A,Z,U,N,M)
0014          CALL PTICR5(U,UO,M,AN2)
0015          CALL PTICR9(AS2O,AN2,Z,N,ALF,RO)
0016          CALL PTICR4(G,U,UO,A,N,M)
0017          CALL PTICR8(G,Z,N,ALF,RO)
0018          CALL PTICR2(H,ODO,N)
       C * starting iterations
0019       14 CONTINUE
0020          ITER=ITER+1
0021       13 ICH=0
       C * ICH - set changed flag
       C * ICH=0 - active constraints not changed
0022          IF(JCH) 1,2,3
0023        1 CALL PTICR2(PLUS,1DO,N)
       C * PLUS(I)=0 - I-th constraint is active
       C * PLUS(I)=1 - I-th constraint isn't active
0024          DO 4 I=1,N
0025          IF(Z(I).GT.O..OR.G(I).LT..O)GOTO 5
0026          PLUS(I)=0
0027          Z(I)=0.
0028        5 CONTINUE
0029        4 CONTINUE
       C * set of active constraints is ready
0030          IDIM=0
0031          DO 15 I=1,N
0032       15 IDIM=IDIM+PLUS(I)+0.5
       C * IDIM - dimension of current face
0033          ICH=1
0034          GOTO 2
0035        3 CONTINUE
```

```
      C * adding new active constraint
0036        PLUS(IPRIM)=0
0037        Z(IPRIM)=0
0038      6 ICH=1
0039        IDIM=IDIM-1
0040      2 CONTINUE
0041        CALL PTICR7(G,G,PLUS,N,ANGRI)
0042        ITERBN=ITERBN+1
      C * ITERBN - face iterations count
0043        IF(ICH.EQ.1)ITERBN=1
0044        IF(ANGRI.GT.ANGRD.AND.ITERBN.NE.
            .IDIM+1.AND.IEND.NE.1)GOTO 7
0045        IF(IEND.EQ.1.AND.IDIM.EQ.IDIMO)GOTO 99
      C * if IEND=1 - test if exact minimum is reached
0046        IEND=1-IEND
0047        IDIMO=IDIM
0048        IF(IEND.EQ.0)GOTO 7
0049        JCH=-1
0050        GOTO 8
0051      7 CONTINUE
      C * getting descent direction - H
0052        BT=0.
0053        IF(ICH.EQ.0)BT=ANGRI/ANGRIO
      C * saving norm of the gradient of the projection
0054        ANGRIO=ANGRI
0055        DO 9 I=1,N
0056      9 H(I)=(BT*H(I)+G(I))*PLUS(I)
      C * descent direction found
      C * Attention!  machine constant
0057        ALMAX=1.E18
0058        IPRIM=0
0059        DO 10 I=1,N
0060        IF(H(I).LE..0)GOTO 11
0061        AL=Z(I)/H(I)
0062        IF(ALMAX.LT.AL)GOTO 12
0063        ALMAX=AL
0064        IPRIM=I
0065     12 CONTINUE
0066     11 CONTINUE
0067     10 CONTINUE
      C * maximum step ALMAX found
      C * IPRIM - number of the new possible active constraint
0068        CALL PTICRO(A,Z,G,U1,H,ALMAX,AL,N,M,ALF,RO,IED)
0069        CALL PTICR1(Z,H,Z,N,-AL)
```

```
0070          CALL PTICR3(A,Z,U,N,M)
0071          CALL PTICR4(G,U,UO,A,N,M)
0072          CALL PTICR8(G,Z,N,ALF,RO)
0073          CALL PTICR7(G,G,PLUS,N,ANGRI)
0074          CALL PTICR5(U,UO,M,AN2)
0075          CALL PTICR9(AS2,AN2,Z,N,ALF,RO)
0076          JCH=0
0077          IF(IED.EQ.1)JCH=1
       C * IED=1 - getting new constraint
0078          IF(IED.EQ.2)GOTO 22
0079          IF(AN2.LE.DL2.OR.ITER.GE.IMAX)GOTO 20
0080          IF(AS2.GE.AS20)GOTO 21
0081          AS20=AS2
0082          GOTO 14
0083    20    CONTINUE
       C * exit by discrepancy level or iteration number
0084          IERR=1
0085          GOTO 999
0086      99 CONTINUE
       C * exact minimum is found
0087          IERR=0
0088          GOTO 999
0089       8 CONTINUE
       C * return to the beginning to check exact minimum
0090          GOTO 13
0091    21    CONTINUE
       C * functional hasn't changed
0092          IERR=2
0093          GOTO 999
0094    22    CONTINUE
       C * negative step found
0095          IERR=65
0096          GOTO 999
0097   999    CONTINUE
0098          RETURN
0099          END


0001          SUBROUTINE PTISR2(A,Z,N,M,IC,S)
       C * transformation of the operator matrix A(M,N)
       C * and initial approximation Z
       C * to the first quadrant
       C * S(N) - work array
       C * IC - correctness set
```

```
0002            IMPLICIT REAL*8(A-H,O-Z)
0003            IMPLICIT INTEGER*4(I-N)
0004            DIMENSION A(M,N),Z(N),S(N)
0005            EXTERNAL PTICR1,PTISR4
0006            IC1=IABS(IC)+1
0007            N1=N-1
0008            GOTO (800,801,802,803,804,805),IC1
0009     800    CONTINUE
       C * non-negative functions
       C * nothing to do
0010            GOTO 999
       C * transformation of the initial point - Z
0011     801    CONTINUE
       C * monotone functions IC=1
0012            DO 101 I=2,N
0013     101    S(I-1)=Z(I-1)-Z(I)
0014            S(N)=Z(N)
0015            GOTO 799
0016     802    CONTINUE
       C * descending, upwards convex functions IC=2
0017            DO 102 I=2,N1
0018     102    S(I)=(I-N)*(Z(I-1)-2.*Z(I)+Z(I+1))
0019            S(N)=Z(N)
0020            S(1)=(Z(1)-Z(2))*(N-1.)
0021            GOTO 799
0022     803    CONTINUE
       C * upwards convex functions IC=3
0023            DO 103 I=2,N1
0024     103    S(I)=(Z(I-1)-2.*Z(I)+
              +Z(I+1))*(N-I)*(I-1.)/(1.-N)
0025            S(N)=Z(N)
0026            S(1)=Z(1)
0027            GOTO 799
0028     804    CONTINUE
       C * descending, downwards convex functions IC=4
0029            DO 104 I=2,N1
0030     104    S(I-1)=(Z(I-1)-2.*Z(I)+Z(I+1))*(I-0.)
0031            S(N1)=(Z(N1)-Z(N))*(N-1.)
0032            S(N)=Z(N)
0033            GOTO 799
0034     805    CONTINUE
       C * downwards convex functions, positive at ends IC=5
0035            DO 105 I=2,N1
0036     105    S(I)=(Z(I-1)-2.*Z(I)+
```

```
                  +Z(I+1))*(N-I)*(I-1.)/(N-1.)
0037              S(N)=Z(N)
0038              S(1)=Z(1)
0039              GOTO 799
0040        799   CONTINUE
0041              CALL PTICR1(S,S,Z,N,ODO)
        C * transformation of matrix A
0042              DO 1 K=1,M
        C * filling array S
0043              CALL PTICR2(S,ODO,N)
0044              DO 2 J=1,N
0045              DO 3 I=1,N
        C * calculating the entries of the transformation matrix
0046              T=PTISR4(I,J,IC,N)
0047        3     S(J)=S(J)+A(K,I)*T
0048        2     CONTINUE
        C * K-th row of matrix A is converted
0049              DO 4 J=1,N
0050        4     A(K,J)=S(J)
0051        1     CONTINUE
0052        999   CONTINUE
0053              RETURN
0054              END


0001              FUNCTION PTISR4(I,J,IC,N)
        C * calculation of the entries of the transformation matrix
        C * from PI-plus to compact, specified by parameter IC
0002              IMPLICIT REAL*8(A-H,O-Z)
0003              IMPLICIT INTEGER*4(I-N)
0004              IC1=IABS(IC)+1
0005              GOTO (800,801,802,803,804,805),IC1
0006        800   CONTINUE
        C * non-negative
0007              H=0.
0008              IF(I.EQ.J)H=1.
0009              GOTO 998
0010        801   CONTINUE
        C * descending functions IC=1
0011              H=0.
0012              IF(I.LE.J)H=1.
0013              GOTO 998
0014        802   CONTINUE
        C * descending, upwards convex functions IC=2
```

```
0015          H=1.
0016          IF(I.GT.J)H=(N-I)/(N-J+0.)
0017          GOTO 998
0018    803   CONTINUE
C * upwards convex IC=3
0019          IF(I.LE.J.AND.J.NE.1)H=(I-1.)/(J-1.)
0020          IF(I.GE.J.AND.J.NE.N)H=(N-I)/(N-J+0.)
0021          GOTO 998
0022    804   CONTINUE
C * descending, downwards convex functions IC=4
0023          H=0.
0024          IF(I.LE.J.AND.J.NE.N)H=(J-I+1.)/(J+0.)
0025          IF(J.EQ.N)H=1.
0026          GOTO 998
0027    805   CONTINUE
C * downwards convex functions, positive at ends IC=5
0028          IF(I.LE.J.AND.J.NE.1)H=(I-1.)/(J-1.)
0029          IF(I.GE.J.AND.J.NE.N)H=(N-I)/(N-J+0.)
0030          IF(J.NE.1.AND.J.NE.N)H=-H
0031          GOTO 998
0032    998   CONTINUE
0033          PTISR4=H
0034          RETURN
0035          END


0001          SUBROUTINE PTISR3(Z,N,IC,Y)
C * transformation from PI-plus to the set specified by IC
0002          IMPLICIT REAL*8(A-H,O-Z)
0003          IMPLICIT INTEGER*4(I-N)
0004          DIMENSION Z(N),Y(N)
0005          EXTERNAL PTISR4,PTICR1
0006          DO 1 I=1,N
0007          S=0.
0008          DO 2 J=1,N
0009    2     S=S+Z(J)*PTISR4(I,J,IC,N)
0010          Y(I)=S
0011    1     CONTINUE
0012          CALL PTICR1(Y,Y,Z,N,ODO)
0013          RETURN
0014          END
```

IX. General programs

```
0001          SUBROUTINE PTICRO(AK,A,X1,X2,Y1,Y2,N,M)
0002          IMPLICIT REAL*8(A-H,O-Z)
0003          IMPLICIT INTEGER*4(I-N)
       C * Gets system of linear equations
       C * for finite-difference approximation of
       C * integral equation of the first kind
       C * AK(X,Y) - kernel function of the integral equation
       C * X1,X2 - integration limits
       C * Y1,Y2 - Y-variable range
       C * N - number of numerical integration points
       C * M - number of points of Y-grid
       C * A(M,N) - matrix of the linear system (output)
       C * Integration over the second variable
       C *      of the function AK(X,Y)
0004          DIMENSION A(M,N)
0005          H=(X2-X1)/(N-1.)
0006          DO 1 I=1,N
0007          X=X1+(X2-X1)/(N-1.)*(I-1.)
0008          DO 2 J=1,M
0009          Y=Y1+(Y2-Y1)/(M-1.)*(J-1.)
0010          A(J,I)=AK(X,Y)*H
0011          IF(I.EQ.1.OR.I.EQ.N)A(J,I)=A(J,I)/2.
0012    2     CONTINUE
0013    1     CONTINUE
0014          RETURN
0015          END


0001          SUBROUTINE PTICRO(A,Z,G,U,H,ALM,AL,N,M,
              *ALF,RO,IED)
       C * one-dimensional minimization
       C *      of the smoothing functional
       C *      on a half-line Z-AL*H AL>0
       C * A(M,N) - matrix of the linear operator
       C * Z(N) - start point
       C * G(N) - gradient at the point Z
       C * H(N) - descent direction
       C * ALM - step restriction AL<ALM
       C * U(M) - work array
       C * AL - step found
       C * ALF - regularization parameter
       C * RO - weight of the difference derivative
```

```
      C *                                    in the norm of W21
      C * IED - return code
      C *   0 - normal - minimum found,
      C *   1 - minimum on the restriction found,
      C *   2 - negative step found
      C *   3 - zero division
0002        IMPLICIT REAL*8(A-H,O-Z)
0003        IMPLICIT INTEGER*4(I-N)
0004        DIMENSION A(M,N),Z(N),G(N),U(M),H(N)
0005        EXTERNAL PTICR6,PTICR3
0006        H2=0.
0007        CALL PTICR6(G,H,N,GH)
0008        IF(GH.LT.0.)GOTO 2
0009        CALL PTICR3(A,H,U,N,M)
0010        CALL PTICR6(U,U,M,AH2)
0011        IF(ALF.EQ.0.)GOTO 3
      C * regularization parameter doesn't equal zero
      C * find norm of the descent direction
0012        DO 4 I=2,N
0013    4   H2=H2+(H(I)-H(I-1))**2*RO+H(I)**2
0014        H2=H2+H(1)**2
0015    3   CONTINUE
0016        IF(AH2+H2*ALF.EQ.0.0)GOTO 998
0017        AL=GH/(AH2+H2*ALF)*.5
0018        IF(AL.GE.ALM)GOTO 1
0019        IED=0
0020        GOTO 999
0021    1   CONTINUE
      C * minimum on the restriction (bound) found
0022        AL=ALM
0023        IED=1
0024        GOTO 999
0025    2   CONTINUE
      C * negative step found, make it equal to zero
0026        AL=0.
0027        IED=2
0028        GOTO 999
0029  998   CONTINUE
      C * zero division, let step be zero
0030        IED=3
0031        AL=0.
0032        GOTO 999
0033  999   CONTINUE
0034        RETURN
```

```
0035          END


0001          SUBROUTINE PTICR1(B,C,A,N,P)
       C * array moving
       C * A(I):=B(I)+P*C(I), I=1,N
0002          IMPLICIT REAL*8(A-H,O-Z)
0003          IMPLICIT INTEGER*4(I-N)
0004          DIMENSION A(N),B(N),C(N)
0005          DO 1 I=1,N
0006     1    A(I)=B(I)+P*C(I)
0007          RETURN
0008          END


0001          SUBROUTINE PTICR2(A,R,N)
       C * fill real array A of length N
       C * by the value R
0002          IMPLICIT REAL*8(A-H,O-Z)
0003          IMPLICIT INTEGER*4(I-N)
0004          DIMENSION A(N)
0005          DO 1 I=1,N
0006     1    A(I)=R
0007          RETURN
0008          END


0001          SUBROUTINE PTICI2(A,R,N)
       C * fill integer array A of length N
       C * by the value R
0002          INTEGER*4 N,I,R,A
0003          DIMENSION A(N)
0004          DO 1 I=1,N
0005     1    A(I)=R
0006          RETURN
0007          END


0001          SUBROUTINE PTICR3(A,Z,U,N,M)
       C * multiply matrix A(M,N) by vector Z(N)
       C * U(M) - vector-result
0002          IMPLICIT REAL*8(A-H,O-Z)
0003          IMPLICIT INTEGER*4(I-N)
0004          DIMENSION U(M),A(M,N),Z(N)
```

```
0005          DO 1 I=1,M
0006          S=0.
0007          DO 2 J=1,N
0008     2    S=S+A(I,J)*Z(J)
0009     1    U(I)=S
0010          RETURN
0011          END
```


```
0001          SUBROUTINE PTICR4(G,U,U0,A,N,M)
       C * calculating gradient of discrepancy (norm AZ-U0)
       C * U=A*Z - value of the operator at the point Z
       C * G(N) - result
0002          IMPLICIT REAL*8(A-H,O-Z)
0003          IMPLICIT INTEGER*4(I-N)
0004          DIMENSION G(N),U(M),U0(M),A(M,N)
0005          DO 1 J=1,N
0006          S=0.
0007          DO 2 I=1,M
0008     2    S=S+A(I,J)*(U(I)-U0(I))
0009     1    G(J)=S*2.
0010          RETURN
0011          END
```


```
0001          SUBROUTINE PTICR5(A,B,N,S)
       C * calculating discrepancy
       C * S - result
0002          IMPLICIT REAL*8(A-H,O-Z)
0003          IMPLICIT INTEGER*4(I-N)
0004          DIMENSION A(N),B(N)
0005          S=0.
0006          DO 1 I=1,N
0007     1    S=S+(A(I)-B(I))**2
0008          RETURN
0009          END
```


```
0001          SUBROUTINE PTICR6(A,B,N,S)
       C * calculating scalar product of vectors A and B
       C * S - result
0002          IMPLICIT REAL*8(A-H,O-Z)
0003          IMPLICIT INTEGER*4(I-N)
0004          DIMENSION A(N),B(N)
```

```
0005          S=0.
0006          DO 1 I=1,N
0007    1     S=S+A(I)*B(I)
0008          RETURN
0009          END


0001          SUBROUTINE PTICR7(A,B,C,N,S)
        C * scalar product of vectors A and B of length N
        C *      with weights C(I)
        C * S - result
0002          IMPLICIT REAL*8(A-H,O-Z)
0003          IMPLICIT INTEGER*4(I-N)
0004          DIMENSION A(N),B(N),C(N)
0005          S=0.
0006          DO 1 I=1,N
0007    1     S=S+A(I)*B(I)*C(I)
0008          RETURN
0009          END


0001          SUBROUTINE PTICR8(G,Z,N,ALF,RO)
        C * calculating the gradient of the stabilizer
        C *      with weight RO
        C *      and adding it to the vector G(N)
        C * Z(N) - point to calculate
        C * ALF - regularization parameter
0002          IMPLICIT REAL*8(A-H,O-Z)
0003          IMPLICIT INTEGER*4(I-N)
0004          DIMENSION Z(N),G(N)
0005          IF(ALF.EQ.0.)GOTO 999
0006          DO 1 I=1,N
0007          G(I)=G(I)+2.*ALF*Z(I)
0008          IF(I.NE.1.AND.I.NE.N)G(I)=G(I)+
              *2.*ALF*(2.*Z(I)-Z(I-1)-Z(I+1))*RO
0009          IF(I.EQ.1)
              +     G(I)=G(I)+2.*ALF*(Z(1)-Z(2))*RO
0010          IF(I.EQ.N)
              +     G(I)=G(I)+2.*ALF*(Z(N)-Z(N-1))*RO
0011    1     CONTINUE
0012    999   CONTINUE
0013          RETURN
0014          END
```

```
0001          SUBROUTINE PTICR9(AS2,AN2,Z,N,ALF,RO)
       C * calculate stabilizer and add it to discrepancy AN2
       C * Z(N) - point
       C * ALF - regularization parameter
       C * RO - weight of the difference derivative
       C *                              in the norm of W21
       C * AS2 - result
0002          IMPLICIT REAL*8(A-H,O-Z)
0003          IMPLICIT INTEGER*4(I-N)
0004          DIMENSION Z(N)
0005          S=0.
0006          IF(ALF.EQ.0.)GOTO 999
0007          DO 4 I=2,N
0008     4    S=S+(Z(I)-Z(I-1))**2*RO+Z(I)**2
0009          S=S+Z(1)**2
0010   999    CONTINUE
0011          AS2=AN2+S*ALF
0012          RETURN
0013          END
```

Postscript

In conclusion we would like to briefly consider the contents of the main monographs on ill-posed problems that have appeared after 1990. We see three main directions of research:

1) variational methods, including Tikhonov's approach with various means for choosing the regularization parameter;
2) iterative methods;
3) methods of statistical regularization.

Of course, in all three directions one studies both linear and nonlinear ill-posed problems.

1. Variational methods

The most complete investigations regarding the application of variational methods for solving nonlinear ill-posed problems is the monograph 'Nonlinear ill-posed problems' by A.N. Tikhonov, A.S. Leonov, and A.G. Yagola. This monograph was written already at the beginning of 1991, and its publication (in Russian by Publishing Company 'Nauka' and in English by Chapman & Hall Publishers) was delayed until 1995. The most general problem regarded in it is the minimization of a functional defined on a topological space. Under certain assumptions, it turns out to be possible to regularize the problem of finding an extremum of the functional, using the generalized functional of Tikhonov and by choosing the regularization parameter in accordance with the generalized discrepancy principle (see Chapter 1 of the present book), with the generalized principle of quasisolutions, and with the generalized principle of the smoothing functional. After this it is shown how to regularize nonlinear operator equations. The following important question is considered in great detail: to numerically solve operator equations using regularizing algorithms, the initial problem has to be approximated by finite-dimensional problems. By strictly underpinning a suitable approach as well as finite-dimensional analogs of the generalized principles, it becomes possible to implement the methods on a computer and to give a practical application of these methods.

Numerous studies in the theory of ill-posed problems have been devoted to generalizing the results expounded in the present book and concerning the solution of ill-posed problems on the set of monotone bounded functions to the space of functions of bounded variation. This problem is also considered in some detail in 'Nonlinear ill-posed problems'. In it, algorithms for solving both linear and nonlinear ill-posed problems under the condition that the exact solution is a function of bounded variation have been given. These algorithms not only guarantee mean-square convergence of the regularized approximations, but also, under certain conditions, uniform convergence.

Methods for solving nonlinear operator equations can be successfully used also in linear algebra, in considering the problem of finding the pseudo-inverse of a matrix with minimal norm (the method of minimal pseudo-inversion of a matrix). As a result one succeeds in constructing a regularizing algorithm for finding the normal pseudosolution of a system of linear algebraic equations, which has certain optimal properties. Certain practical applications are also considered in 'Nonlinear ill-posed problems'; in particular, to the inverse problem of vibrational spectroscopy. This problem is considered in more detail in the monograph 'Inverse problems of vibrational spectroscopy' by I.V. Kochikov, G.M. Kuramshina, Yu.A. Pentin, and A.G. Yagola (Moscow Univ. Publ., 1993), which is devoted to the development of regularizing algorithms for constructing multi-atom molecular force fields from experimental data, obtained mainly by the method of vibrational spectroscopy, the possibility of using, in the solution of inverse problems, nonempirical quantum-mechanical calculations, program libraries and the results of data processing for series of molecules.

2. Iterative methods

One of the most interesting directions in the theory of regularizing algorithms is that of iterative methods for solving ill-posed problems. The monograph of A.B. Bakushinsky and A.V. Goncharsky, 'Iterative methods for solving ill-posed problems' (published by 'Nauka' in Russian in 1989 and, in an extended and reworked English version, by Kluwer Acad. Publ. in 1994 under the title 'Ill-posed problems: Theory and applications') stands out among the most complete and interesting publications in this direction.

In that book a general approach towards the construction of regularizing algorithms is considered, using the scheme of approximating a discontinuous map A^{-1} by a family of continuous maps R_α. Tikhonov's approach clearly falls under this scheme, as do most iterative methods.

In the book general problems typical for solving ill-posed problems are given a major place. Such are the problem of regularizability, i.e. the problem of the existence of regularizing algorithms, the problem of estimating the error in the solution at a point or on a set, and the problem of constructing optimal (in a sense) regularizing algorithms.

Within the general framework of constructing regularizing algorithms a broad spectrum of iterative processes is studied and fundamentally new ideas have been proposed for constructing approximate solutions. A large number of mathematical problems has been studied, such as the problem of solving linear or nonlinear operator equations, the problem of minimizing functionals in a Banach space, the problem of finding saddle points and equilibrium points, the linear programming problem, etc.

The efficiency of the methods is illustrated by the solution of numerous applied problems. Among the most interesting such problems are: the problem of image processing, the inverse problem of geophysics, and problems of computerized tomography within linear and nonlinear models. The problem of computerized tomography can be formulated as the inverse coefficient problem for second-order partial differential equations.

Various regularizing algorithms for these problems have been studied. Their efficiency is illustrated by the solution of model and actual problems. The methods worked out lie at the basis of a mathematical software package for computing diagnostics, ensuring a break-through technology in the study of objects on a micro scale, in engineering seismics, etc.

Among the books addressing the latter issues we single out the book by V.V. Vasin and A.L. Ageev, 'Ill-posed problems with a priori information' (Ekaterinburg, Ural Publ. Firm 'Nauka', 1993), in which both general questions on the construction of variational and iterative regularizing algorithms (including Prox's method and Fejer processes), a number of concrete ill-posed problems (regularization of symmetric spectral problems, moment problems), as well as problems on the discrete approximation of regularizing algorithms are considered.

3. Statistical methods

A break-through in the domain of developing statistical methods for ill-posed problems has been the appearance of the monograph 'Ill-posed problems with random noise in the data' by A.M. Fedotov (Novosibirsk, 'Nauka', 1990), with as predecessor 'Linear ill-posed problems with random noise in the data' by the same author (Novosibirsk, 'Nauka', 1982). Using the mathematical model of inverse problems, adequately describing actual physical experiments, the author subsequently uses the theory of weak distributions to consider the problem of statistical correctness of the formulation of linear and nonlinear ill-posed problems, the existence and numerical construction of optimal methods, and the problem of finite-dimensional approximations of the solution. Various classes of solution procedures are considered, including the minimax and Bayesian procedures. The book is distinguished by its high theoretical level, its strict mathematical formulations, and its theoretical substantiation of statistical regularization methods.

We will now turn to several other publications.

4. Textbooks

Next to the well-known textbooks on ill-posed problems by A.N. Tikhonov and V.Ya. Arsenin [78] and C.W. Groetsch [212], during the last years textbooks of various levels of complexity have appeared. First of all, there is the book by A.B. Bakushinsky and A.V. Goncharsky, 'Ill-posed problems: Theory and applications' (Moscow Univ. Publ., 1989) (the material of this book lies at the basis of the above-mentioned English version published by Kluwer Acad. Publ. in 1994), which is intended for students of the Faculty of computational mathematics and cybernetics of Moscow University.

The book by G.M. Wing, 'A primer on integral equations of the first kind: th problem of deconvolution and unfolding' (SIAM, 1991) is in the category of simpl textbooks. The following are in the category of more complicated textbooks: J Baumeister, 'Stable solutions of inverse problems' (Vieweg, 1987); A.K. Louis, 'Invers und schlicht gestellte Probleme' (Teubner, 1989); C.W. Groetsch, 'Inverse problem in the mathematical sciences' (Vieweg, 1993).

5. Handbooks and Conference Proceedings

The most complete handbook, incorporating in our views the majority of direction of theoretical research in the domain of ill-posed problems and its applications is the book published after the International Conference on 'Ill-posed problems in the natural sciences' (Moscow, 19–25 August 1991). A book with this title has been published in English under the redaction of A.N. Tikhonov concurrently by TVP in Moscow and VSP BV in Holland. The proceedings of this conference are interesting because on the basis of them a sufficiently detailed bibliography on ill-posed problems and their applications can be compiled.

Bibliography

[1] O.M. Alifanov, E.A. Artyukhin, S.V. Rumyantsev, *Extremal methods for solving ill-posed problems*, Nauka, Moscow, 1988. (In Russian.)

[2] V.A. Antonyuk, E.I. Rau, D.O. Savin, et al., Recovering the spatial distribution of two-dimensional microfields by methods of reconstructive computational tomography, *Izv. AN SSSR Ser. Fiz.* **51: 3** (1987), pp. 475–479. (In Russian.)

[3] V.Ya. Arsenin, V.V. Ivanov, Solving certain first kind integral equations of convolution type by the regularization method, *Zh. Vychisl. Mat. i Mat. Fiz.* **8: 2** (1968), pp. 310–321. (In Russian.)

[4] O.B. Arushanyan, The discipline of programming in the standard language FORTRAN, *Questions of constructing program libraries* MGU, Moscow, (1980), pp. 3–18. (In Russian.)

[5] A.V. Baev, *Choosing the regularization parameter for operators specified with an error*, MGU, Moscow, 1976. Deposited at VINITI, nr. 493–76. (In Russian.)

[6] A.V. Baev, *Solving an inverse seismological problem*, Moscow, 1977. Dissertation Cand. Phys.-Math. Sc. (In Russian.)

[7] A.V. Baev, Constructing a normal solution of nonlinear ill-posed problems by the regularization method, *Zh. Vychisl. Mat. i Mat. Fiz.* **19: 3** (1979), pp. 594–600. (In Russian.)

[8] A.B. Bakushinsky, A numerical method for solving Fredholm integral equations of the first kind, *Zh. Vychisl. Mat. i Mat. Fiz.* **5: 4** (1965), pp. 744–749. (In Russian.)

[9] A.B. Bakushinsky, A general method for constructing regularizing algorithms for ill-posed linear equations in Hilbert space, *Zh. Vychisl. Mat. i Mat. Fiz.* **7: 3** (1967), pp. 672–676. (In Russian.)

[10] A.B. Bakushinsky, On the extension of the discrepancy principle, *Zh. Vychisl. Mat. i Mat. Fiz.* **10: 1** (1970), pp. 210–213. (In Russian.)

[11] A.B. Bakushinsky, Regularizing algorithms for solving ill-posed extremal problems, *Methods for controlling large systems* Izd. IGU, Irkutsk, **1** (1970), pp. 223–235. (In Russian.)

[12] A.B. Bakushinsky, Remarks on a class of regularizing algorithms, *Zh. Vychisl. Mat. i Mat. Fiz.* **13: 6** (1973), pp. 1596–1598. (In Russian.)

[13] A.B. Bakushinsky, Substantiating the discrepancy principle, *Differential and integral equations* Izd. IGU, Irkutsk, (1973), pp. 117–126. (In Russian.)

[14] A.B. Bakushinsky, A regularizing algorithm based on the Newton–Kantorovich method for solving variational inequalities, *Zh. Vychisl. Mat. i Mat. Fiz.* **16: 6** (1976), pp. 1397–1404. (In Russian.)

[15] A.B. Bakushinsky, Methods for solving monotone variational inequalities based on the iterative regularization principle, *Zh. Vychisl. Mat. i Mat. Fiz.* **17: 6** (1977), pp. 1350–1362. (In Russian.)

[16] A.B. Bakushinsky, A.V. Goncharsky, *Iteration methods for solving ill-posed problems*, Nauka, Moscow, 1988. (In Russian.)

[17] N.S. Bakhvalov, *Numerical methods*, Nauka, Moscow, 1975. (In Russian.)

[18] G.S. Bisnovaty-Kogan, A.V. Goncharsky, B.V. Komberg, et al., Optical characteristics of an accretive disc in the Her system (Her X-I), *Astr. Zh.* **54: 2** (1977), pp. 241–253. (In Russian.)

[19] M.B. Bogdanov, ˙ Finding the brightness distribution of a star based on an analysis of photometric observations of it when covered by the Moon, *Astr. Zh.* **55: 3** (1978), pp. 490–495. (In Russian.)

[20] M.B. Bogdanov, The brightness distribution of the carbonide star TX Pisces obtained by an analysis of photometric observations of it when covered by the Moon, *Astr. Zh.* **56: 5** (1979), pp. 1023–1029. (In Russian.)

[21] M.B. Bogdanov, Finding a source function from observations of the darkening on the boundary of a star, *Astr. Zh.* **57: 2** (1980), pp. 296–301. (In Russian.)

[22] M.B. Bogdanov, Applying the regularization method to the analysis of close double stars and radiosources covered by the Moon, *Astr. Zh.* **57: 4** (1980), pp. 762–766. (In Russian.)

[23] S. Bochner, *Vorlesungen über Fouriersche Integrale*, Chelsea, reprint, 1948.

[24] B.M. Budak, E.M. Berkovich, The approximation of extremal problems. 1, *Zh. Vychisl. Mat. i Mat. Fiz.* **11: 3** (1971), pp. 580–596. (In Russian.)

[25] B.M. Budak, E.M. Berkovich, The approximation of extremal problems. 2, *Zh. Vychisl. Mat. i Mat. Fiz.* **11: 4** (1971), pp. 870–884. (In Russian.)

[26] B.M. Budak, E.M. Berkovich, Yu.L. Gaponenko, The construction of strongly converging minimizing sequences for a continuous convex functional, *Zh. Vychisl. Mat. i Mat. Fiz.* **9: 2** (1969), pp. 286–299. (In Russian.)

[27] B.M. Budak, F.P. Vasil'ev, *Approximate methods for solving optimal control problems*, **2** VTs. MGU, Moscow, 1969. (In Russian.)

[28] B.M. Budak, A. Vin'oli, Yu.L. Gaponenko, A regularization means for a continuous convex functional, *Zh. Vychisl. Mat. i Mat. Fiz.* **9: 5** (1969), pp. 1046–1056. (In Russian.)

[29] M.M. Vainberg, *Functional analysis*, Prosvesh., Moscow, 1979. (In Russian.)

[30] G.M. Vainikko, Error estimates for the method of successive approximation in the case of ill-posed problems, *AiT* **3** (1980), pp. 84–92. (In Russian.)

[31] G.M. Vainikko, A.Yu. Veretennikov, *Iteration procedures in ill-posed problems*, Nauka, Moscow, 1980. (In Russian.)

[32] F.P. Vasil'ev, *Numerical methods for solving extremal problems*, Nauka, Moscow, 1980. (In Russian.)

[33] F.P. Vasil'ev, *Methods for solving extremal problems*, Nauka, Moscow, 1981. (In Russian.)

[34] V.V. Vasin, V.P. Tatana, Approximate solution of operator equations of the first kind, *Mat. Zap. Uralsk. Univ.* **6: 2** (1968), pp. 27–37. (In Russian.)

[35] E.S. Ventsel', V.G. Kobylinsky, A.M. Levin, Application of the regularization method for the numerical solution of the problem of bending of thin elastic plates, *Zh. Vychisl. Mat. i Mat. Fiz.* **24: 2** (1984), pp. 323–328. (In Russian.)

[36] V.A. Vinokurov, The error of the solution of linear operator equations, *Zh. Vychisl. Mat. i Mat. Fiz.* **10: 4** (1970), pp. 830–839. (In Russian.)

[37] V.A. Vinokurov, General properties of the error of the approximate solution of linear functional equations, *Zh. Vychisl. Mat. i Mat. Fiz.* **11: 1** (1971), pp. 22–28. (In Russian.)

[38] V.A. Vinokurov, Two remarks on choosing the regularization parameter, *Zh. Vychisl. Mat. i Mat. Fiz.* **12: 2** (1972), pp. 481–483. (In Russian.)

[39] V.A. Vinokurov, The error order of computing functions with approximately given argument, *Zh. Vychisl. Mat. i Mat. Fiz.* **13: 5** (1973), pp. 1112–1123. (In Russian.)

[40] V.V. Voevodin, *Numerical methods in linear algebra*, Nauka, Moscow, 1966. (In Russian.)

[41] V.V. Voevodin, On the regularization method, *Zh. Vychisl. Mat. i Mat. Fiz.* **9: 3** (1969), pp. 671–673. (In Russian.)

[42] Yu.L. Gaponenko, The method of successive approximation for solving nonlinear extremal problems, *Izv. Vuzov. Mat.* **5** (1980), pp. 12–16. (In Russian.)

[43] F. Gill, U. Murray (eds.), *Numerical methods in conditional optimization*, Mir, Moscow, 1977. (Translated from the English.) (In Russian.)

[44] S.F. Gilyazov, Stability of the solution of linear operator equations of the first kind by the method of fast descent, *Vestn. MGU Ser. 15* **3** (1980), pp. 26–32. (In Russian.)

[45] S.F. Gilyazov, *Methods for solving linear ill-posed problems*, Izd. MGU, Moscow, 1987. (In Russian.)

[46] V.B. Glasko, G.V. Gushchin, V.I. Starostenko, The method of Tikhonov regularization for solving systems of nonlinear equations, *DAN USSR B* **3** (1975), pp. 203–206. (In Russian.)

[47] V.B. Glasko, G.V. Gushchin, V.I. Starostenko, Applying the regularization method of A.N. Tikhonov for solving nonlinear systems of equations, *Zh. Vychisl. Mat. i Mat. Fiz.* **16: 2** (1976), pp. 283–292. (In Russian.)

[48] N.L. Goldman, Approximate solution of Fredholm integral equations of the first kind in the class of piecewise-convex functions with bounded first derivative, *Numerical analysis in FORTRAN. Methods and algorithms* MGU, Moscow, (1978), pp. 30–40. (Translated from the English.) (In Russian.)

[49] A.V. Goncharsky, Numerical methods for solving ill-posed problems on compact sets, *All-Union Conf. on Ill-posed Problems* Frunze, 1979, p. 47. (In Russian.)

[50] A.V. Goncharsky, Inverse problems of optics, *Vestn. MGU Ser. 15* **3** (1986) pp. 59–77. (In Russian.)

[51] A.V. Goncharsky, Ill-posed problem and methods for solving them, *Ill-posed problems in the natural sciences* MGU, Moscow, 1987, pp. 15–36. (In Russian.)

[52] A.V. Goncharsky, Mathematical models in problems of synthesis of flat optical elements, *Computer optics* **1** MTsNTI, Moscow, 1987. (In Russian.)

[53] A.V. Goncharsky, G.V. Gushin, A.G. Yagola, A numerical experiment for solving two-dimensional Fredholm integral equations of the first kind, *Certain problems in the automation of processing interpretations of physical experiments* **1** MGU, Moscow, 1973. pp. 192–201. (In Russian.)

[54] A.V. Goncharsky, G.V. Gushchin, A.S. Leonov, et al., Certain algorithms for searching the approximate solution of ill-posed problems in the set of monotone functions, *Zh. Vychisl. Mat. i Mat. Fiz.* **12: 2** (1972), pp. 283–297. (In Russian.)

[55] A.V. Goncharsky, A.V. Kolpakov, A.A. Stepanov, Inverse problems of numerical diagnostics of damaged surfaces of layers of crystals with Röntgendiffractional data, *Surfaces* **12** (1986), pp. 67–71. (In Russian.)

[56] A.V. Goncharsky, A.S. Leonov, A.G. Yagola, The solution of two-dimensional Fredholm equations of the first kind with a kernel depending on the difference of the arguments, *Zh. Vychisl. Mat. i Mat. Fiz.* **11: 5** (1971), pp. 1296–1301. (In Russian.)

[57] A.V. Goncharsky, A.S. Leonov, A.G. Yagola, A generalized discrepancy principle for the case of operators specified with an error, *DAN SSSR* **203: 6** (1972), pp. 1238–1239. (In Russian.)

[58] A.V. Goncharsky, A.S. Leonov, A.G. Yagola, A regularizing algorithm for ill-posed problems with approximately specified operator, *Zh. Vychisl. Mat. i Mat. Fiz.* **12: 6** (1972), pp. 1592–1594. (In Russian.)

[59] A.V. Goncharsky, A.S. Leonov, A.G. Yagola, A generalized discrepancy principle, *Zh. Vychisl. Mat. i Mat. Fiz.* **13: 2** (1973), pp. 294–302. (In Russian.)

[60] A.V. Goncharsky, A.S. Leonov, A.G. Yagola, Methods for solving Fredholm integral equations of the first kind of convolution type, *Certain problems in the automation of processing and interpreting physical experiments* MGU, Moscow, **1** (1973), pp. 170–191. (In Russian.)

[61] A.V. Goncharsky, A.S. Leonov, A.G. Yagola, Finite-difference approximation of linear ill-posed problems, *Zh. Vychisl. Mat. i Mat. Fiz.* **14: 1** (1974), pp. 15–24. (In Russian.)

[62] A.V. Goncharsky, A.S. Leonov, A.G. Yagola, The discrepancy principle for solving nonlinear ill-posed problems, *DAN SSSR* **214: 3** (1974), pp. 499–500. (In Russian.)

[63] A.V. Goncharsky, A.S. Leonov, A.G. Yagola, The solution of ill-posed problems with approximately specified operator, *Proc. All-Union school of young scientists: 'Methods for solving ill-posed problems, and their application'* MGU, Moscow, (1974), pp. 39–43. (In Russian.)

[64] A.V. Goncharsky, A.S. Leonov, A.G. Yagola, The applicability of the discrepancy principle in the case of nonlinear ill-posed problems and a new regularizing algorithm for solving them, *Zh. Vychisl. Mat. i Mat. Fiz.* **15: 2** (1975), pp. 290–297. (In Russian.)

[65] A.V. Goncharsky, V.V. Stepanov, Algorithms for approximately solving ill-posed problems on certain compact sets, *DAN SSSR* **245: 6** (1979), pp. 1296–1299. (In Russian.)

[66] A.V. Goncharsky, V.V. Stepanov, Uniform approximation of solutions with bounded variation for ill-posed problems, *DAN SSSR* **248: 1** (1979), pp. 20–22. (In Russian.)

[67] A.V. Goncharsky, V.V. Stepanov, Numerical methods for solving ill-posed problems on compact sets, *Vestnik MGU, Ser. 15* **3** (1980), pp. 12–18. (In Russian.)

[68] A.V. Goncharsky, V.V. Stepanov, Inverse problems of coherent optics. Focussing in line, *Zh. Vychisl. Mat. i Mat. Fiz.* **26: 1** (1986), pp. 80–91. (In Russian.)

[69] A.V. Goncharsky, S.M. Tivkov, G.S. Khromov, A.G. Yagola, Studies in the spatial structure of planetary nebulae by the method of mathematical modelling, *Pis'ma v Astron. Zh.* **5: 8** (1979), pp. 406–410. (In Russian.)

[70] A.V. Goncharsky, A.M. Cherepashchuk, A.G. Yagola, Application of the regularization method for solving inverse problems in astrophysics, *Proc. All-Union school of young scientists: 'Methods for solving ill-posed problems, and their application'* MGU, Moscow, (1974), pp. 21–24. (In Russian.)

[71] A.V. Goncharsky, A.M. Cherepashchuk, A.G. Yagola, *Numerical methods for solving inverse problems in astrophysics*, Nauka, Moscow, 1978. (In Russian.)

[72] A.V. Goncharsky, A.M. Cherepashchuk, A.G. Yagola, Physical characteristics of Wolf–Raye stars and Röntgen sources in close binary systems, *All-Union conf. on ill-posed problems.* Frunze, (1979), p. 48. (In Russian.)

[73] A.V. Goncharsky, A.M. Cherepashchuk, A.G. Yagola, *Ill-posed problems problems in astrophysics*, Nauka, Moscow, 1985. (In Russian.)

[74] A.V. Goncharsky, A.G. Yagola, Uniform approximation of the monotone solution of ill-posed problems, *DAN SSSR* **184: 4** (1969), pp. 771–773. (In Russian.)

[75] A.V. Goncharsky, A.G. Yagola, Solving integral equations of the form $\int_a^b K(x,s)\,dg(s) = u(x)$, *DAN SSSR* **193: 2** (1970), pp. 266–267. (In Russian.)

[76] V.F. Dem'yanov, A.M. Rubinov, *Approximate methods for solving extremal problems*, LGU, Leningrad, 1968. (In Russian.)

[77] T.F. Dolgopolova, V.K. Ivanov, Numerical differentiation, *Zh. Vychisl. Mat. i Mat. Fiz.* **6: 3** (1966), pp. 570–575. (In Russian.)

[78] I.N. Dombrovskaya, Equations of the first kind with closed operator, *Izv. Vuzov. Mat.* **6** (1967), pp. 39–72. (In Russian.)

[79] I.N. Dombrovskaya, V.K. Ivanov, On the theory of linear ill-posed equations in abstract spaces, *Sibirsk. Mat. Zh.* **6: 3** (1965), pp. 499–508. (In Russian.)

[80] I.V. Emelin, M.A. Krasnosel'sky, Stop rule in iterative procedures for solving ill-posed problems, *AiT* **12** (1978), pp. 59–63. (In Russian.)

[81] I.V. Emelin, M.A. Krasnosel'sky, On the theory of ill-posed problems, *DAN SSSR* **244: 4** (1979), pp. 805–808. (In Russian.)

[82] P.N. Zaikin, The numerical solution of inverse problems of operational computations in the real domain, *Zh. Vychisl. Mat. i Mat. Fiz.* **8: 2** (1968), pp. 411–415. (In Russian.)

[83] V.K. Ivanov, A uniqueness theorem for the inverse problem of the logarithmic potential for starlike sets, *Izv. Vuzov. Mat.* **3** (1958), pp. 99–106. (In Russian.)

[84] V.K. Ivanov, Stability for the inverse problem of the logarithmic potential for starlike sets, *Izv. Vuzov. Mat.* **4** (1958), pp. 96–99. (In Russian.)

[85] V.K. Ivanov, Integral equations of the first kind and the approximate solution of the inverse problem for the logarithmic potential, *DAN SSSR* **142: 5** (1962), pp. 997–1000. (In Russian.)

[86] V.K. Ivanov, Linear ill-posed problems, *DAN SSSR* **145: 5** (1962), pp. 270–272. (In Russian.)

[87] V.K. Ivanov, Ill-posed problems, *Mat. Sb.* **61: 2** (1963), pp. 211–223. (In Russian.)

[88] V.K. Ivanov, The approximate solution of operator equations of the first kind, *Zh. Vychisl. Mat. i Mat. Fiz.* **6: 6** (1966), pp. 1089–1094. (In Russian.)

[89] V.K. Ivanov, Ill-posed problems in topological spaces, *Sibirsk. Mat. Zh.* **10: 5** (1969), pp. 1065–1074. (In Russian.)

[90] V.K. Ivanov, V.V. Vasin, V.P. Tanana, *The theory of linear ill-posed problems and its applications*, Nauka, Moscow, 1978. (In Russian.)

[91] V.A. Il'in, E.G. Poznyak, *Fundamentals of mathematical analysis*, **1** Nauka, Moscow, 1982. (In Russian.)

[92] K. Yosida, *Functional analysis*, Springer, Berlin, 1980.

[93] V.G. Karmanov, *Mathematical programming*, Nauka, Moscow, 1986. (In Russian.)

[94] K.N. Kasparov, A.S. Leonov, A photo-emission method for measuring spectra and a new means for mathematical processing them, *Zh. Prikl. Spektroskopi.* **22: 3** (1975), pp. 491–498. (In Russian.)

[95] A.N. Kolmogorov, S.V. Fomin, *Elements of the theory of functions and functional analysis*, Nauka, Moscow, 1989. (In Russian.)

[96] A.L. Korotaev, P.I. Kuznetsov, A.S. Leonov, V.G. Rodoman, Determining the optimal regime for introducing an anti-bacterial preparate, *AiT* **1** (1986), pp. 100–106. (In Russian.)

[97] I.V. Kochikov, G.M. Kuramshina, Yu.A. Pentin, A.G. Yagola, A regularizing algorithm for solving the inverse vibrational problem, *DAN SSSR* **261: 5** (1981), pp. 1104–1106. (In Russian.)

[98] I.V. Kochikov, G.M. Kuramshina, Yu.A. Pentin, A.G. Yagola, Computation of the force field of a multi-atomic molecule by the regularization method, *DAN SSSR* **283: 4** (1985), pp. 850–854. (In Russian.)

[99] I.V. Kochikov, G.M. Kuramshina, A.G. Yagola, Stable numerical methods for

solving certain inverse problems of vibrational spectroscopy, *Zh. Vychisl. Mat. i Mat. Fiz.* **27: 11** (1987), pp. 350–357. (In Russian.)

[100] I.V. Kochikov, A.N. Matvienko, A.G. Yagola, A modification of the generalized discrepancy principle, *Zh. Vychisl. Mat. i Mat. Fiz.* **23: 16** (1983), pp. 1298–1303. (In Russian.)

[101] I.V. Kochikov, A.N. Matvienko, A.G. Yagola, The generalized discrepancy principle for solving incompatible problems, *Zh. Vychisl. Mat. i Mat. Fiz.* **24: 7** (1983), pp. 1087–1090. (In Russian.)

[102] I.V. Kochikov, A.N. Matvienko, A.G. Yagola, The regularization method for solving nonlinear incompatible equations, *Zh. Vychisl. Mat. i Mat. Fiz.* **27: 3** (1987), pp. 456–458. (In Russian.)

[103] I.V. Kochikov, V.D. Rusov, M.Yu. Semenov, A.G. Yagola, Strengthening weak images in an autoradiographical experiment, *Integral equations in applied modelling* Kiev Univ., Kiev, (1983), pp. 166–167. (In Russian.)

[104] M.M. Lavrent'ev, The Cauchy problem for the Laplace equation, *Izv. AN SSSR* **20** (1956), pp. 819–842. (In Russian.)

[105] M.M. Lavrent'ev, The question of the inverse problem in potential theory, *DAN SSSR* **106: 3** (1956), pp. 389–390. (In Russian.)

[106] M.M. Lavrent'ev, The Cauchy problem for linear elliptic equations of the second order, *DAN SSSR* **112: 2** (1957), pp. 195–197. (In Russian.)

[107] M.M. Lavrent'ev, Integral equations of the first kind, *DAN SSSR* **127: 1** (1959), pp. 31–33. (In Russian.)

[108] M.M. Lavrent'ev, *Certain ill-posed problems of mathematical physics*, SO AN SSSR, Novosibirsk, 1962. (In Russian.)

[109] M.M. Lavrent'ev, V.G. Romanov, S.P. Shishatsky, *Ill-posed problems of mathematical physics and analysis*, Nauka, Moscow, 1980. (In Russian.)

[110] A.M. Levin, Regularizing the computation of the infima of functionals, *Zh. Vychisl. Mat. i Mat. Fiz.* **24: 8** (1984), pp. 1123–1128. (In Russian.)

[111] A.M. Levin, Computing the incompatibility measure of operator equations of the first kind, *Zh. Vychisl. Mat. i Mat. Fiz.* **26: 4** (1986), pp. 499–507. (In Russian.)

[112] A.S. Leonov, Constructing stable difference schemes for solving nonlinear boundary value problems, *DAN SSSR* **244: 3** (1975), pp. 525–528. (In Russian.)

[113] A.S. Leonov, Stable solution of the inverse problem in gravimetry on the class of convex bodies, *Izv. AN SSSR. Fiz. Zeml.* **7** (1976), pp. 55–65. (In Russian.)

[114] A.S. Leonov, Algorithms for the approximate solution of nonlinear ill-posed problems with approximately specified operator, *DAN SSSR* **245: 2** (1979), pp. 300–304. (In Russian.)

[115] A.S. Leonov, Choosing the regularization parameter for nonlinear ill-posed problems with approximately specified operator, *Zh. Vychisl. Mat. i Mat. Fiz.* **19: 6** (1979), pp. 1363–1376. (In Russian.)

[116] A.S. Leonov, Relation between the generalized discrepancy method and the generalized discrepancy principle for nonlinear ill-posed problems, *Zh. Vychisl.*

Mat. i Mat. Fiz. **22**: **4** (1982), pp. 783–790. (In Russian.)

[117] A.S. Leonov, Certain algorithms for solving ill-posed extremal problems, *Mat. Sb.* ' **129 (171)**: **2** (1986), pp. 218–231. (In Russian.)

[118] A.S. Leonov, M.S. Suleĭmanova, A method for processing the results of dynamic radionuclide studies on the basis of regularization methods for solving ill-posed problems, *Med. Radiologiya* **12** (1985), pp. 60–63. (In Russian.)

[119] O.A. Liskovets, Regularization of equations with closed linear operator, *DAN BSSR* **15**: **6** (1971), pp. 481–483. (In Russian.)

[120] O.A. Liskovets, Optimal properties of the principle of the smoothing functional, *DAN SSSR* **234**: **2** (1977), pp. 298–301. (In Russian.)

[121] O.A. Liskovets, A projection method for realizing the discrepancy method for nonlinear equations of the first kind with perturbed operator, *Diff. Uravnen.* **16**: **4** (1980), pp. 723–731. (In Russian.)

[122] O.A. Liskovets, *Variational methods for solving unstable problems*, Nauka i Tekhn., Minsk, 1981. (In Russian.)

[123] L.A. Lyusternik, V.I. Sobolev, *Elements of functional analysis*, Nauka, Moscow, 1965. (In Russian.)

[124] D.Ya. Martynov, A.V. Goncharsky, A.M. Cherepashchuk, A.G. Yagola, Application of the regularization method to the solution of the inverse problem of astrophysics, *Probl. Mat. Fiz. i Vychisl. mat.* Nauka, Moscow, (1971), pp. 205–218. (In Russian.)

[125] G.I. Marchuk, *Methods of numerical mathematics*, Nauka, Moscow, 1989. (In Russian.)

[126] S.N. Mergelyan, Harmonic approximation and approximate solution of the Cauchy problem for the Laplace equation, *UMN* **11**: **5** (1956), pp. 3–26. (In Russian.)

[127] V.A. Morozov, The solution of functional equations by the regularization method, DAN SSSR, **167**: **3** (1966), pp. 510–512. (In Russian.)

[128] V.A. Morozov, Regularization of ill-posed problems and choice of the regularization parameter, *Zh. Vychisl. Mat. i Mat. Fiz.* **6**: **1** (1966), pp. 170–175. (In Russian.)

[129] V.A. Morozov, Regularizing families of operators, *Computing methods and programming* **8** MGU, Moscow, (1967), pp. 63–93. (In Russian.)

[130] V.A. Morozov, Choice of the parameter when solving functional equations by the regularization method, *DAN SSSR* **175**: **6** (1967), pp. 1225–1228. (In Russian.)

[131] V.A. Morozov, The discrepancy principle for solving operator equations by the regularization method, *Zh. Vychisl. Mat. i Mat. Fiz.* **8**: **2** (1968), pp. 295–309. (In Russian.)

[132] V.A. Morozov, Regularization of a class of of extremum problems, *Computing methods and programming* **12** MGU, Moscow, (1969), pp. 24–37. (In Russian.)

[133] V.A. Morozov, The problem of differentiation and certain algorithms for approximating experimental data, *Computing methods and programming* **14** MGU, Moscow, (1970), pp. 45–62. (In Russian.)

[134] V.A. Morozov, A new approach to the solution of linear equations of the first kind with approximate operator, *Proc. 1st conf. young scientists of the faculty of VMiK MGU* MGU, Moscow, (1973), pp. 22–28. (In Russian.)

[135] V.A. Morozov, The discrepancy principle for solving incompatible equations by A.N. Tikhonov's regularization method, *Zh. Vychisl. Mat. i Mat. Fiz.* **13: 5** (1973), pp. 1099–1111. (In Russian.)

[136] V.A. Morozov, V.I. Gordonova, Numerical algorithms for choosing the parameter in the regularization method, *Zh. Vychisl. Mat. i Mat. Fiz.* **13: 3** (1973), pp. 539–545. (In Russian.)

[137] V.A. Morozov, Computing the infima of functionals from approximate information, *Zh. Vychisl. Mat. i Mat. Fiz.* **13: 4** (1973), pp. 1045–1048. (In Russian.)

[138] V.A. Morozov, *Regular methods for solving ill-posed problems*, Nauka, Moscow, 1987. (In Russian.)

[139] V.A. Morozov, An optimality principle for the discrepancy when approximately solving equations with nonlinear operator, *Zh. Vychisl. Mat. i Mat. Fiz.* **14: 4** (1974), pp. 819–827. (In Russian.)

[140] V.A. Morozov, Regular methods for solving nonlinear operator equations, *Izv. Vuzov. Mat.* **11** (1978), pp. 74–86. (In Russian.)

[141] V.A. Morozov, N.L. Gol'dman, M.K. Samarin, Descriptive regularization method and quality of approximation, *Inzh. Fiz. Zh.* **33: 6** (1977), pp. 1117–1124. (In Russian.)

[142] V.A. Morozov, N.N. Kirsanova, A.F. Sysoev, A library of algoritms for the fast Fourier transformation for discrete series, *Numerical analysis in FORTRAN* MGU, Moscow, 1976. (In Russian.)

[143] A.Ya. Perel'man, V.A. Punina, Application of the Mellin convolution to the solution of integral equations of the first kind with kernel depending on the product, *Zh. Vychisl. Mat. i Mat. Fiz.* **9: 3** (1969), pp. 626–646. (In Russian.)

[144] E. Polak, *Numerical optimization methods*, Mir, Moscow, 1974. (Translated from the English.) (In Russian.)

[145] B.T. Polyak, Minimization methods in the presence of constraints, *Itogi Nauk. i Tekhn. Mat. Anal.* **12** VINITI, Moscow, 1974. (In Russian.)

[146] B.N. Pshenichny, Yu.M. Danilin, *Numerical methods in extremum problems*, Nauka, Moscow, 1975. (In Russian.)

[147] I.M. Rapoport, The planar inverse problem of potential theory, *DAN SSSR* **28** (1940), pp. 305–307. (In Russian.)

[148] I.M. Rapoport, Stability in the inverse problem of potential theory, *DAN SSSR* **31** (1941), pp. 305–307. (In Russian.)

[149] M.K. Samarin, Stability of the solution of certain ill-posed problems on sets of special structure, *Dissertation Cand. Phys.-Math. Sc.* 1979. (In Russian.)

[150] M.K. Samarin, Mean square approximation by piecewise monotone functions, *Numerical analysis in FORTRAN* **15** MGU, Moscow, (1976), pp. 83–90. (In Russian.)

[151] A.A. Samarsky, *Theory of difference schemes*, Nauka, Moscow, 1989. (In Russian.)

[152] V.K. Saul'ev, I.I. Samoĭlov, Approximate methods for unconditional optimization of functions of several variables, *Itogi Nauk. i Tekhn. Mat. Anal.* VINITI, Moscow, **11** (1973), pp. 91–128. (In Russian.)

[153] S.L. Sobolev, *Some applications of functional analysis in mathematical physics*, LGU, Leningrad, 1950. (In Russian.)

[154] V.I. Starostenko, *Stable numerical methods in gravimetric problems*, Nauk. Dumka, Kiev, 1978. (In Russian.)

[155] V.V. Stepanov, Numerical solution of ill-posed problems on the set of piecewise monotone and of convex functions, *Vestnik MGU Ser. 15* **3** (1985), pp. 21–26. (In Russian.)

[156] V.N. Strakhov, The solution of ill-posed problems of magneto- and gravimetry representable by integral equations of convolution type. 1, *Izv. AN SSSR. Fiz. Zeml.* **4** (1967), pp. 36–54. (In Russian.)

[157] V.N. Strakhov, The solution of ill-posed problems of magneto- and gravimetry representable by integral equations of convolution type. 2, *Izv. AN SSSR. Fiz. Zeml.* **5** (1967), pp. 33–53. (In Russian.)

[158] V.N. Strakhov, The numerical solution of ill-posed problems representable by integral equations of convolution type, *DAN SSSR* **178: 2** (1968), pp. 299–302. (In Russian.)

[159] Yu.N. Syshchikov, L.V. Vilkov, A.G. Yagola, A method for determining the potential of internal rotation functions from electrongraphical data, *Vestnik MGU. Khimiya* **24: 6** (1983), pp. 541–543. (In Russian.)

[160] V.P. Tanana, Ill-posed problems and the geometry of Banach spaces, *DAN SSSR* **193: 1** (1970), pp. 43–45. (In Russian.)

[161] V.P. Tanana, Approximate solution of operator equations of the first kind and geometric properties of Banach spaces, *Izv. Vuzov. Mat.* **7** (1971), pp. 81–93. (In Russian.)

[162] V.P. Tanana, A projection-iterative algorithm for operator equations of the first kind with perturbed operator, *DAN SSSR* **224: 5** (1975), pp. 1028–1029. (In Russian.)

[163] V.P. Tanana, *Methods for solving operator equations*, Nauka, Moscow, 1981. (In Russian.)

[164] A.N. Tikhonov, Stability of inverse problems, *DAN SSSR* **39: 5** (1943), pp. 195–198. (In Russian.)

[165] A.N. Tikhonov, The solution of ill-posed problems and the regularization method, *DAN SSSR* **151: 3** (1963), pp. 501–504. (In Russian.)

[166] A.N. Tikhonov, Regularization of ill-posed problems, *DAN SSSR* **153: 1** (1963), pp. 49–52. (In Russian.)

[167] A.N. Tikhonov, Solving nonlinear integral equations of the first kind, *DAN SSSR* **56: 6** (1964), pp. 1296–1299. (In Russian.)

[168] A.N. Tikhonov, On nonlinear equations of the first kind, *DAN SSSR* **161: 5** (1965), pp. 1023–1026. (In Russian.)

[169] A.N. Tikhonov, Stability of algorithms for solving degenerate systems of linear algebraic equations, *Zh. Vychisl. Mat. i Mat. Fiz.* **5: 4** (1965), pp. 718–722. (In Russian.)

[170] A.N. Tikhonov, Ill-posed problems of linear algebra and stable methods for solving them, *DAN SSSR* **163: 3** (1965), pp. 591–594. (In Russian.)

[171] A.N. Tikhonov, Ill-posed problems of optimal planning and stable methods for solving them, *DAN SSSR* **164: 3** (1965), pp. 507–510. (In Russian.)

[172] A.N. Tikhonov, Ill-posed problems of optimal planning, *Zh. Vychisl. Mat. i Mat. Fiz.* **6: 1** (1966), pp. 81–89. (In Russian.)

[173] A.N. Tikhonov, Stability of the problem of optimizing functionals, *Zh. Vychisl. Mat. i Mat. Fiz.* **6: 4** (1966), pp. 631–634. (In Russian.)

[174] A.N. Tikhonov, A reciprocity principle, *DAN SSSR* **253: 2** (1980), pp. 305–308. (In Russian.)

[175] A.N. Tikhonov, Normal solutions of approximate systems of linear algebraic equations, *DAN SSSR* **254: 3** (1980), pp. 549–554. (In Russian.)

[176] A.N. Tikhonov, Approximate systems of linear algebraic equations, *Zh. Vychisl. Mat. i Mat. Fiz.* **20: 6** (1980), pp. 1373–1383. (In Russian.)

[177] A.N. Tikhonov, Problems with imprecisely given initial information, *DAN SSSR* **280: 3** (1985), pp. 559–563. (In Russian.)

[178] A.N. Tikhonov, V.Ya. Arsenin, *Methods for solving ill-posed problems*, Nauka, Moscow, 1986. (In Russian.)

[179] A.N. Tikhonov, F.P. Vasil'ev, *Methods for solving ill-posed extremum problems*, Banach Centre Publ., 1976. pp. 297–342. (In Russian.)

[180] A.N. Tikhonov, V.Ya. Galkin, P.N. Zaikin, Direct methods for solving optimal control problems, *Zh. Vychisl. Mat. i Mat. Fiz.* **7: 2** (1967), pp. 416–423. (In Russian.)

[181] A.N. Tikhonov, V.B. Glasko, Approximate solution of Fredholm integral equations of the first kind, *Zh. Vychisl. Mat. i Mat. Fiz.* **4: 3** (1964), pp. 564–571. (In Russian.)

[182] A.N. Tikhonov, V.B. Glasko, Application of the regularization method in nonlinear problems, *Zh. Vychisl. Mat. i Mat. Fiz.* **5: 5** (1965), pp. 463–473. (In Russian.)

[183] A.N. Tikhonov, A.V. Goncharsky, I.V. Kochikov, et al., Numerical microtomography of objects in Röntgen and optical scanning microscopy, *DAN SSSR* **289: 5** (1986), pp. 1104–1107. (In Russian.)

[184] A.N. Tikhonov, A.V. Goncharsky, V.V. Stepanov, I.V. Kochikov, Ill-posed problem in image processing, *DAN SSSR* **294: 4** (1987), pp. 832–837. (In Russian.)

[185] A.N. Tikhonov, A.V. Goncharsky, V.V. Stepanov, A.G. Yagola, *Regularizing algorithms and a priori information*, Nauka, Moscow, 1983. (In Russian.)

[186] A.N. Tikhonov, V.G. Karmanov, T.L. Rudneva, Stability of linear programming problems, *Numerical methods and programming* **12** MGU, Moscow, (1969), pp. 3–9. (In Russian.)

[187] V.N. Trushnikov, A stable nonlinear iterative process, *Izv. Vuzov. Mat.* **5** (1979), pp. 58–62. (In Russian.)

[188] V.N. Trushnikov, A nonlinear regularizing algorithm and some applications of it, *Zh. Vychisl. Mat. i Mat. Fiz.* **19: 4** (1979), pp. 822–829. (In Russian.)

[189] J. Wilkinson, C. Rainsch, *Handbook of algorithms in the language ALGOL. Linear Algebra*, Mashinostroenie, Moscow, 1976. (Translated from the English.) (In Russian.)

[190] A.M. Fedotov, *Linear ill-posed problems with random errors in the data*, Nauka, Novosibirsk, 1982. (In Russian.)

[191] A. Fiacco, G. MacCormick, *Nonlinear programming. Methods for unconditional sequential optimization*, Mir, Moscow, 1972. (Translated from the English.) (In Russian.)

[192] J. Himmelblau, *Applied nonlinear programming*, Mir, Moscow, 1975. (Translated from the English.) (In Russian.)

[193] T.S. Huang, *Fast algorithms in numerical image processing*, Radio and Connections, Moscow, 1984. (Translated from the English.) (In Russian.)

[194] A.M. Cherepashchuk, A.V. Goncharsky, A.G. Yagola, Interpretation of brightness curves of the system V 444 Cygni in the continuum, *Astron. Tsirkulyar* **413** (1967), pp. 5–8. (In Russian.)

[195] A.M. Cherepashchuk, A.V. Goncharsky, A.G. Yagola, Interpretation of ecliptic systems as the inverse problem of photometry, *Astron. Zh.* **44: 6** (1967), pp. 1239–1252. (In Russian.)

[196] A.M. Cherepashchuk, A.V. Goncharsky, A.G. Yagola, Interpretation of brightness curves of ecliptic systems under the condition that the unknown function is monotone, *Astron. Zh.* **45: 6** (1968), pp. 1191–1206. (In Russian.)

[197] A.M. Cherepashchuk, A.V. Goncharsky, A.G. Yagola, Stratification of illumination in the envelope of ecliptic-double stars of Wolf–Raye type 444 Cygni, *Astron. Zh.* **49: 3** (1972), pp. 533–543. (In Russian.)

[198] A.M. Cherepashchuk, A.V. Goncharsky, A.G. Yagola, Algorithms and computer programs for solving brightness curves of ecliptic systems containing a component with extensive atmosphere, *Peremenn. Zvezd.* **18: 6** (1973), pp. 535–569. (In Russian.)

[199] A.M. Cherepashchuk, A.V. Goncharsky, A.G. Yagola, Solving brightness curves of ecliptic variable stars in the class of monotone functions, *Astronom. Zh.* **51: 4** (1974), pp. 776–781. (In Russian.)

[200] A.M. Cherepashchuk, A.V. Goncharsky, A.G. Yagola, Interpretations of brightness curves of ecliptic systems with disc-like envelopes. The Hercules HZ system, *Astronom. Zh.* **54: 5** (1977), pp. 1027–1035. (In Russian.)

[201] V.F. Sholokhovich, Unstable extremum problems and geometric properties of Banach spaces, *DAN SSSR* **195: 2** (1970), pp. 289–291. (In Russian.)

[202] A.G. Yagola, Choice of the regularization parameter using the generalized discrepancy principle, *DAN SSSR* **245: 1** (1979), pp. 37–39. (In Russian.)

[203] A.G. Yagola, The generalized discrepancy principle in reflexive spaces, *DAN SSSR* **249: 1** (1979), pp. 71–73. (In Russian.)

[204] A.G. Yagola, Solving nonlinear ill-posed problems using the generalized discrepancy principle, *DAN SSSR* **252: 4** (1980), pp. 810–813. (In Russian.)

[205] A.G. Yagola, Choice of the regularization parameter when solving ill-posed problems in reflexive spaces, *Zh. Vychisl. Mat. i Mat. Fiz.* **20: 3** (1980), pp. 586–596. (In Russian.)

[206] A.G. Yagola, Ill-posed problems with approximately given operator, *Banach Centre Publ.* **13** (1984), pp. 265–279. (In Russian.)

[207] J. Cooley, J. Tuckey, An algorithm for machine calculation of complex Fourier series, *Math. of Comp.* **19: 90** (1965), pp. 297–301.

[208] M. Frank, P. Wolfe, An algorithm for quadratic programming, *Naval Res. Los. Quart.* **3** (1956), pp. 95–110.

[209] M. Furi, A. Vignoli, On the regularization of nonlinear ill-posed problems in Banach spaces, *J. Optim. Theory and Appl.* **4: 3** (1969), pp. 206–209.

[210] C.F. Gauss, *Theoria motus corporum collestrium in sectionibus conicis solem ambientirm*, Hamburg, 1809.

[211] A.V. Goncharsky, V.V. Stepanov, V.L. Khrokhlova, A.G. Yagola, The study of local characteristics of the Ap-stars surfaces, *Publ. Czechoslovak. Akad. Sci., Prague* **54** (1978), pp. 36–37.

[212] C.W. Groetsch, *The theory of Tikhonov regularization for Fredholm equations of the first kind*, Pitman, Boston, 1984.

[213] J. Hadamard, *Le problème de Cauchy et les équations aux dérivées partièlles linéaires hyperbolique*, Hermann, Paris, 1932.

[214] B. Hofmann, *Regularization for applied inverse and ill-posed problems*, Teubner, Leipzig, 1986.

[215] F. John, A note on 'improper' problems in partial differential equations, *Commun. Pure and Appl. Math.* **8** (1955), pp. 591–594.

[216] F. John, Continuous dependence on data for solutions with a prescribed bound, *Commun. Pure and Appl. Math.* **13: 4** (1960), pp. 551–585.

[217] A.M. Legendre, *Nouvelles méthodes pour la détermination dse orbites des cometes*, Courcier, Paris, 1806.

[218] D.L. Phillips, A technique for numerical solution of certain integral equations of the first kind, *J. Assoc. Comput. Machinery* **9: 1** (1962), pp. 84–97.

[219] C. Pucci, Sui problemi Cauchy non 'ben posti', *Atti Acad. Naz. Zincci. Rend. Cl. Sci. Fis. Mat. e Natur.* **18: 5** (1955), pp. 473–477.

[220] C. Pucci, Discussione del problema di Cauchy pour le equazioni di tipo elliptico, *Ann. Mat. Pura ad Appl.* **46** (1958), pp. 131–154.

Index